# Transport and Energy 2021



Held at the 2021 AIChE Annual Meeting

Boston, Massachusetts, USA and Online
7 - 11 November and 15 - 19 November 2021

ISBN: 978-1-7138-5703-7

**Printed from e-media with permission by:**

Curran Associates, Inc.
57 Morehouse Lane
Red Hook, NY 12571



Printed with permission by Curran Associates, Inc. (2022)

For permission requests, please contact AIChE
at the address below.

AIChE
120 Wall Street, FL 23
New York, NY 10005-4020

Phone: (800) 242-4363
Fax: (203) 775-5177

www.aiche.org

**Additional copies of this publication are available from:**

Curran Associates, Inc.
57 Morehouse Lane
Red Hook, NY 12571 USA
Phone: 845-758-0400
Fax: 845-758-2633
Email: curran@proceedings.com
Web: www.proceedings.com

# TABLE OF CONTENTS

# 622660 Towards a New Understanding of the Retro-Aldol Reaction for Oxidative Conversion of Lignin to Aromatic Aldehydes and Acids

**Abstract:**

Lignin is the most abundant natural biopolymer after cellulose and accounts for 30% of non-fossil carbon on earth. The retro-aldol reaction is one of the key steps involved in the oxidative conversion of lignin to aromatic aldehydes and acids. In principle, the retro-aldol reaction can proceed in the absence of oxygen. In this work, a new approach based on the influence of oxygen on the oxidation of lignin was investigated. The effect of reaction chemistry, time, temperature, and lignin feedstock play a key role on the yield of aromatic aldehydes and acids. At 140°C, oxidation of softwood Lignoboost kraft lignin for 40 minutes results in combined maximum yield of 5.17% w/w of vanillin and vanillic acid. In comparison, using the new approach in which oxygen was charged for only 20 minutes during the 40 minute reaction improved this yield considerably to 6.95%. Further, yield improvement was obtained when applying this approach to different lignin feedstocks. Oxidation also increased the carboxyl content of lignin from 0.49 mmol/g to 1.41 mmol/g which represents consequential improvement. The current study provides further evidence showing that the oxidation reaction is a crucial pathway for lignin valorization.

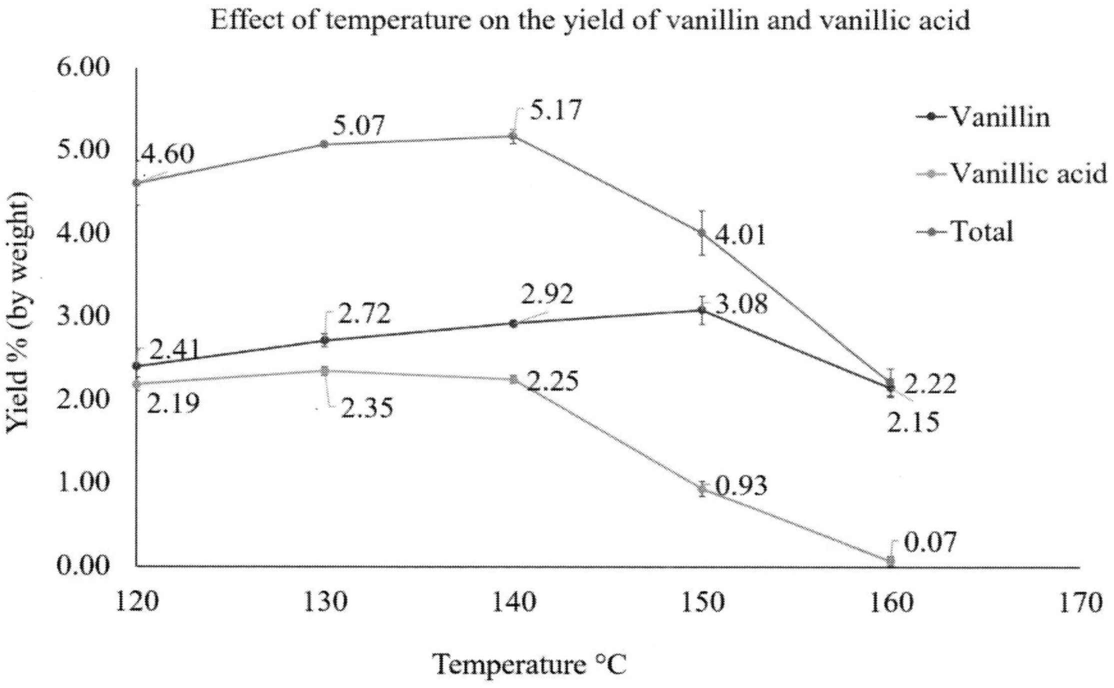


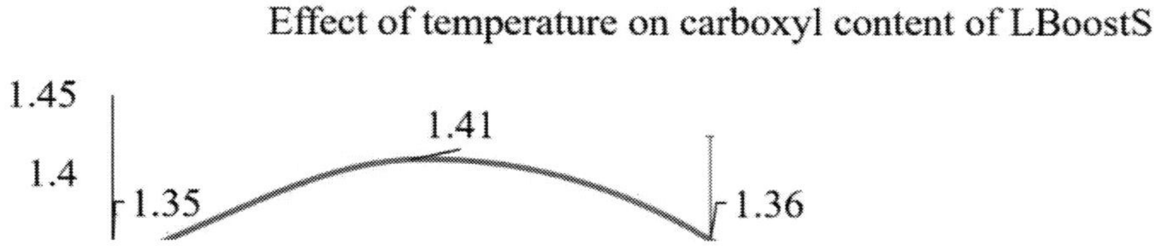

*** Formatting Issue - Best Available Paper/Graphic ***

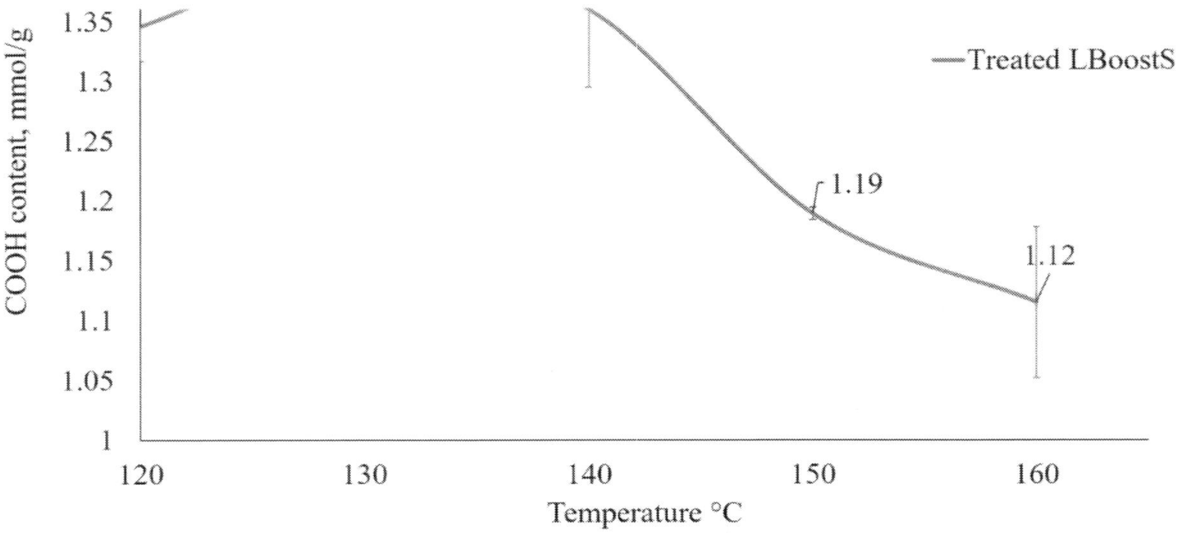


Effect of time on the yield of vanillin and vanillic acid

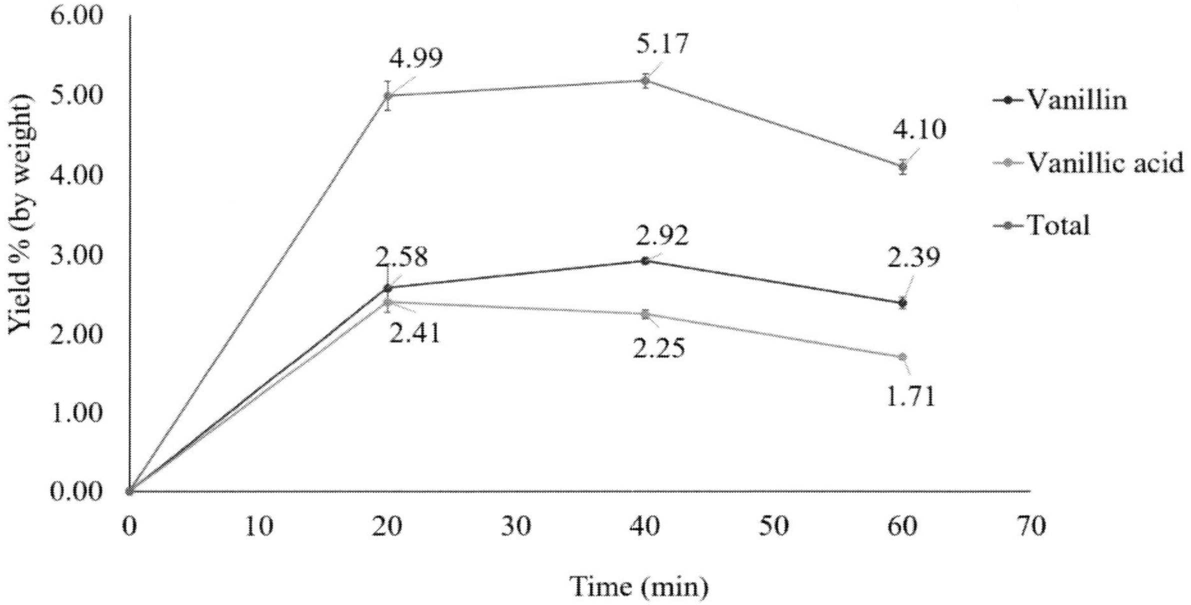


Effect of controlling the amount of oxygen charged during the reaction on the yield of vanillin and vanillic acid

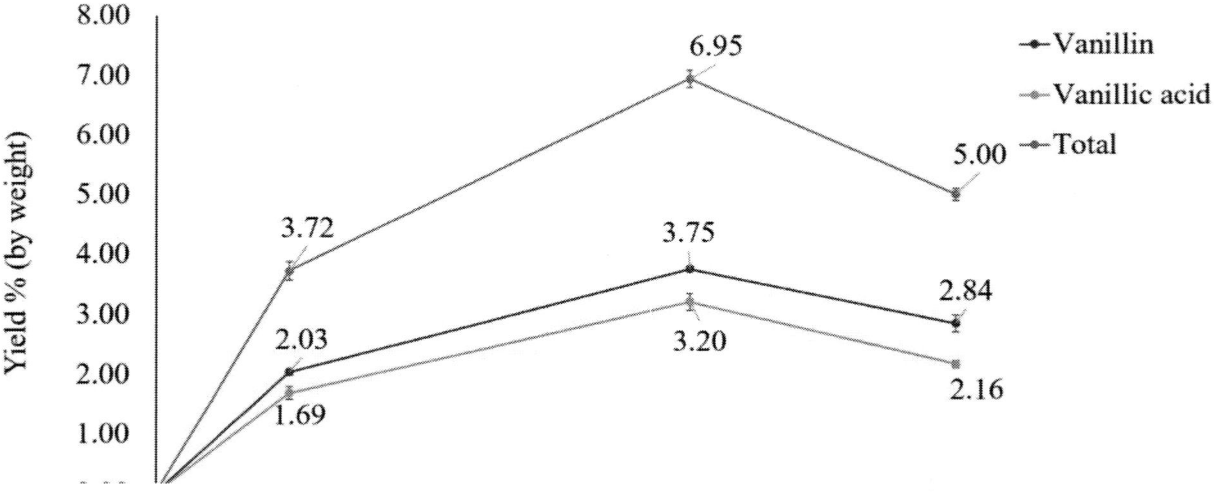

*** Formatting Issue - Best Available Paper/Graphic ***

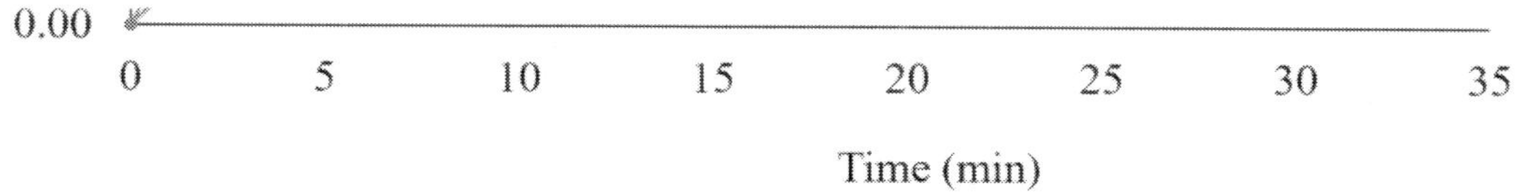


Effect of controlling the amount of oxygen charged during the reaction on the yield of vanillin and vanillic acid from different lignin feedstocks

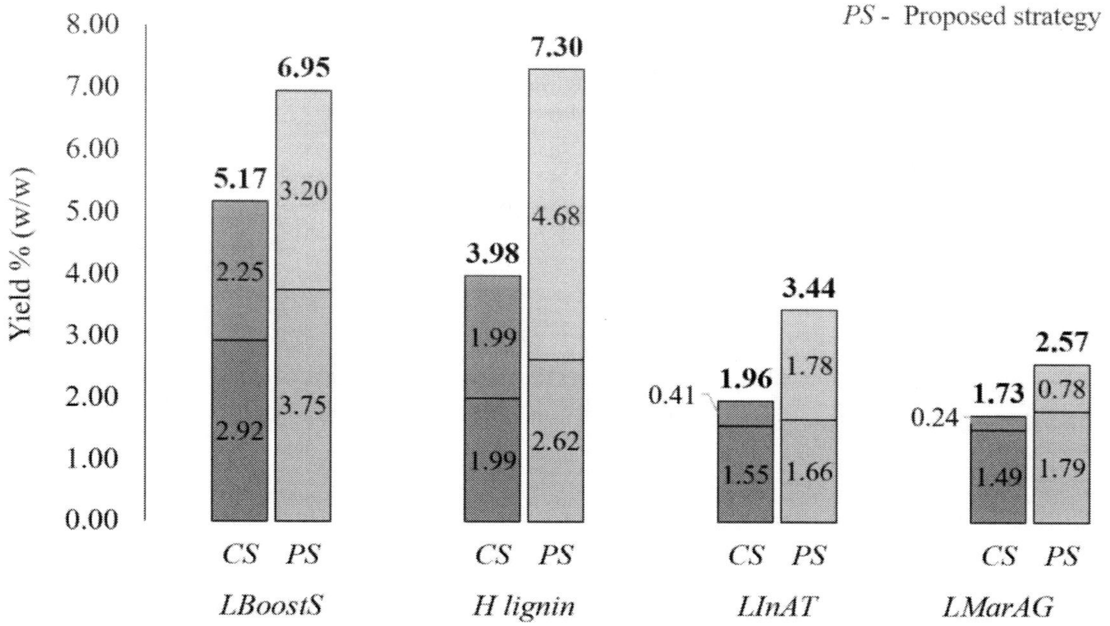

# 628460 Lignin Depolymerization in a Molten Salt Hydrate (lithium bromide trihydrate)

**Abstract:**

A molten salt hydrates (MSH) is a concentrated aqueous solution of an inorganic salt, in which the molar ratio of water to salt is close to the coordination number of the salt cation. Because of their unique properties, in particular the ability to dissolve cellulose, some MSHs are promising solvents and reaction media for biomass processing. For example, LiBr molten salt hydrate (~61 wt% LiBr aqueous solution) has been successfully used in hydrolysis of cellulose and lignocellulosic biomass, lignin quantitation, isomerization of glucose, furan-based chemicals from biomass, synthesis of oligosaccharides, and fabrication of nanocrystals, films, and aerogels from cellulose and biomass in our previous studies. It was also observed that lignin was extensively depolymerized when biomass was treated in the LiBr system. This presentation is to report our recent investigations into the depolymerization mechanisms of lignin in the LiBr system. The studies revealed that all the ether bonds of lignin in methoxyl, β-*O*-4, β-5, and β-β structures except for the 4-*O*-5 bond in diphenyl structure could be cleaved in the LiBr system. The ether bonds were cleaved via the $S_N2$ substitution pathway except for β-*O*-4 bond. The β-*O*-4 bond was primarily cleaved via the benzyl cation and enol ether intermediates, leading to Hibbert's ketones. Besides, some β-*O*-4 structures were transformed into benzodioxane structures, which were stable in the LiBr system.

**Model compounds** **Corresponding structures in lignin**

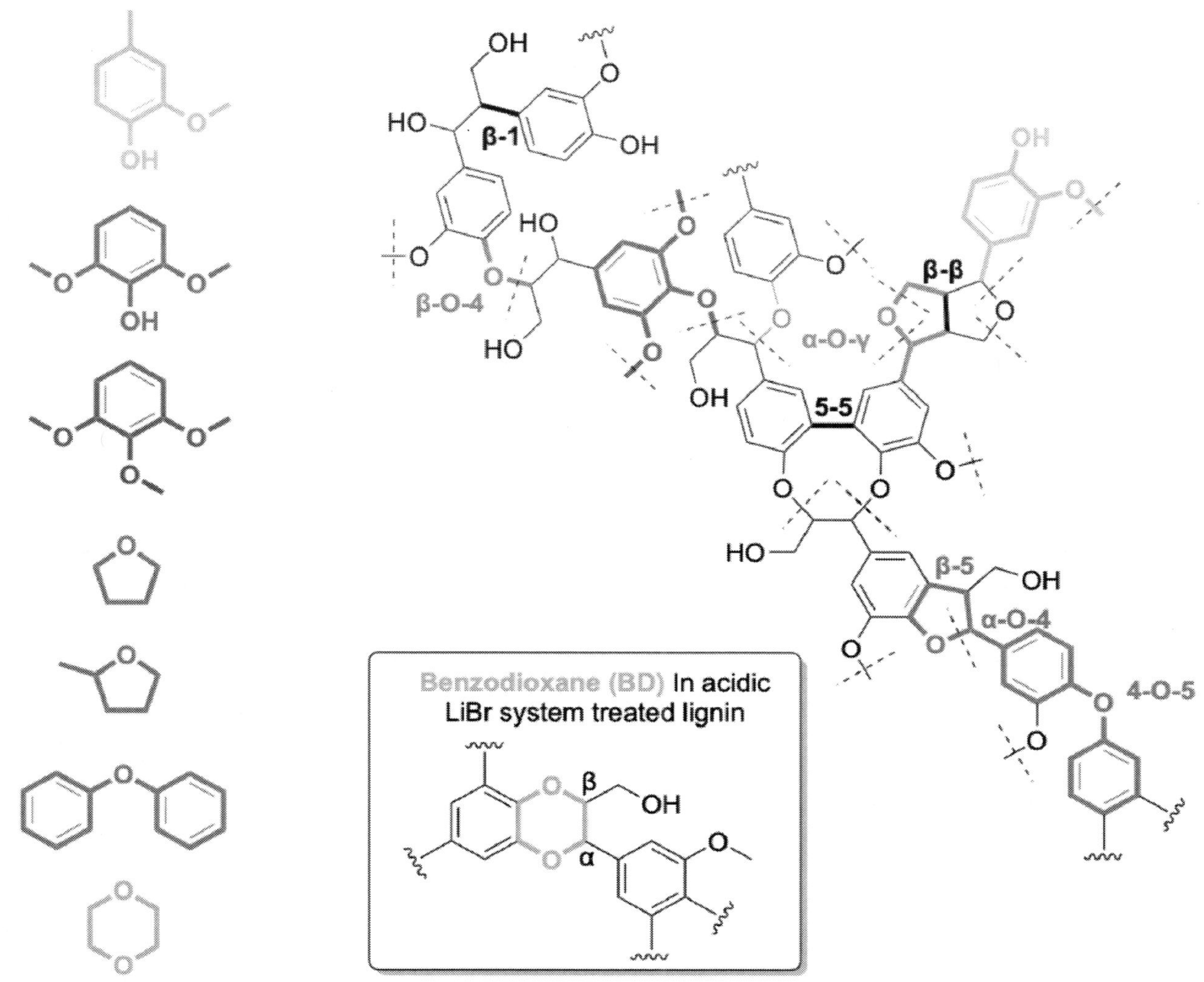

# 631322 Direct Catalytic Conversion of Lignocellulosic Biomass into Liquid Paraffins and Aromatics

**Abstract:**

Today, gasoline is the major economic driver for petroleum refineries and its production accounts for nearly 20% of the total volume of products derived from a barrel of crude oil. However, the market share occupied by gasoline consumption is expected to decline 1% annually through 2050 and the petrochemical industry will become more dependent on the production of non-combustible petrochemicals, jet fuel and marine diesel due to increased car fuel efficiency, emerging electric and hydrogen fuel cell vehicle utilization. Thus, according to the BP Energy Outlook 2020, biofuels are expected to play a major role for the aviation and marine industries, but they also can play a role in accelerating the displacement of petroleum-based gasoline and non-combustible fuels, which could be substituted by the exact bio-based equivalents if there are biomass conversion technologies that enable such potential. Herein, a zeolite-supported metal catalyst in a presence of a hydrocarbon solvent was used to directly convert lignocellulosic biomass into a range of gaseous and liquid hydrocarbons at 280 Celsius via hydrodeoxygenation. Biomass conversion and carbon yields higher than 90% were obtained with a solvent to biomass ratio of 3:1 and 8h of residence time. C5-C7 alkanes (61% of total carbon) and C3-C4 alkanes (14% of total carbon) along with lower level of aromatics (10% of total carbon) were the main products generated. The produced alkanes can be fractionated and used directly in many applications today or subjected to further oxidative coupling followed by oligomerization to produce carbon range jet fuel and diesel.

# Experimental Evaluation of Nonthermal Plasma Reactors for Hydrogen Production from Polyethylene

**Benard Tabu[1], Kevin Akers[1], Peng Yu[2], Wong Hsi-Wu[2], and Juan Pablo Trelles[1]**

[1] Department of Mechanical Engineering, University of Massachusetts, Lowell, MA, United States of America
[2] Department of Chemical Engineering, University of Massachusetts, Lowell, MA, United States of America

## Abstract

Hydrogen is a clean fuel that can also be utilized for in-situ production of water (in combination with atmospheric oxygen). Water and hydrogen-rich low-molecular weight substances, such as methane and ammonia, have been traditionally used as hydrogen sources or hydrogen-storage media. Solids such as plastics, biomass, and solid wastes constitute alternative, abundant, and under-utilized hydrogen sources. Hydrogen production from solids have largely relied on thermochemical approaches, generally operating at relatively high temperatures (e.g., 200 to 700 °C for pyrolysis, 400 to 1500 °C for gasification) and/or pressures (1 to 10s atm). Nonthermal plasmas are partially ionized gases in which the free electrons have significantly larger (equivalent) temperatures than the heavy-species (molecules, atoms, ions), e.g. ~ 1 eV (~11600 K) for electrons compared to a few hundred °C for the heavy-species. Nonthermal plasma approaches have the potential to be more energy efficient and selective means for hydrogen production from solids compared to established thermochemical approaches, especially when operated at low temperature (e.g., < 200 °C) and atmospheric pressure conditions. Moreover, the direct use of electricity to generate chemical reactivity in plasma-based hydrogen production approaches prompts them to be especially environmentally benign when powered by renewable energy sources, such as solar photovoltaic and wind power. Two plasma reactors, one based on a transfer arc (transarc) discharge and the other based on a gliding arc (glidarc) discharge, have been designed, fabricated, and experimentally evaluated for the production of hydrogen from low-density polyethylene (LDPE). The transarc reactor is based on a pin-to-plate electrodes configuration leading to the transfer of electric current through the solid feedstock. The transarc allows direct control of power delivered to the feedstock and requires minimal auxiliary gas flow. The glidarc is based on a three-equally spaced diverging electrodes configuration, leading to the generation of a tri-prong arc that is electrically detached from the feedstock. The arc forms near the base of the electrodes, at the location of minimum electrodes separation, and then glides along the electrodes due to the combined effects of advection by the inflow gas and buoyancy of the low-density plasma, to finally impinge on the solid feedstock. The glidarc allows the processing of arbitrary and heterogeneous feedstock streams, but its operation is sensitive to the inflow gas rate. Both reactors were used for the treatment of solid LDPE feedstock samples in batch mode under atmospheric temperature and pressure conditions. The experiments included time-dependent measurements of power consumed by the plasma, hydrogen production rate, hydrogen production selectivity, and hydrogen production efficiency as function of driving voltage and flow rate. The results show that hydrogen production rate increases with increasing power levels and decreasing flow rates for the glidarc; whereas for the transarc, the hydrogen production rate significantly depends on power level, indistinctively of the flow rate. The results indicate that nonthermal atmospheric plasma approaches are effective for the production of hydrogen from polyethene at mild temperature and pressure conditions.

**Acknowledgements**

This work has been supported by the US Army Combat Capabilities Development Command (CCDC) Soldier Center Contracting Division through Contract # W911QY-20-2-0005. The authors gratefully acknowledge the collaboration with and feedback from Chris Drew and Eric Brack (US Army CDCC) during the execution of this work.

# 633511 Advanced Carbon Capture System for Power Plants

**Abstract:**

TDA Research, Inc (TDA) is developing a high capacity $CO_2$ adsorbent and demonstrating its technical and economic viability for $CO_2$ capture from coal-fired power plants. TDA uses an advanced physical adsorbent to selectively remove $CO_2$ from the flue gas. The sorbent exhibits a much higher affinity for $CO_2$ than $N_2$, $H_2O$ or $O_2$. The sorbent binds $CO_2$ more strongly than common adsorbents, providing the chemical potential needed to remove the $CO_2$. However, because $CO_2$ does not form a true covalent bond with the surface sites, the regeneration can be carried out with only a very small energy input (<10 kcal per mol of $CO_2$) compared to amines (14.2 kcal/mol $CO_2$). We demonstrated the sorbent's efficacy through multiple cycles for removing $CO_2$ from coal derived flue gas and the sorbent demonstrated a stable capacity. The presence of acid gases such as $SO_2$ and NOx (80 and 200 ppmv, respectively) and water vapor up to 15% vol. had no adverse effect on the $CO_2$ capacity. The regeneration of the sorbent and the recovery of $CO_2$ and its pressurization can be achieved by several approaches, including temperature and vacuum swing. A detailed techno-economic analysis using Aspen Plus$^{TM}$ software was conducted in collaboration with University of California Irvine to estimate the impact of the TDA's $CO_2$ capture system on plant efficiency and cost of electricity. Our system analysis showed that TDA's $CO_2$ capture process will reduce plant efficiency by only 19.4% (compared to 31.2% reduction for amine scrubbers) while providing 90% capture. We will present the latest results from the field testing of the unit under the DOE sponsored work at the AIChE meeting.

# 626832 $CO_2$ Absorption in an Aqueous Amine with a Physical Co-Solvent: Electrolyte NRTL and Rate-Based Process Model and Pilot Plant Experimental Validation.

**Abstract:**

Absorption with aqueous amines is the most mature industrial technology to remove $CO_2$ from a gas. The major drawback of the technology is the high energy required to regenerate the solvent. To lower the energy required to warm up the solvent during the regeneration it has been proposed to add a significant amount of a physical co-solvent, to lower the solvent specific heat. There is, however, no clear experimental and model-based evaluation of the overall process of $CO_2$ removal with a mixed aqueous amine and physical co-solvent. This is important to assess the overall impact of the physical co-solvent on the $CO_2$ absorption and desorption properties as well as on the operation.

In this work we present the results of a rate-based AspenPlus process model for the $CO_2$ capture using aqueous mixtures of N-methyl diethanolamine (MDEA) and a common physical co-solvent. We performed new Liquid-Vapor Equilibrium as well as new kinetic experiments and used the data to develop a new thermodynamic electrolyte e-NRTL model, as well as a new kinetic model, which have subsequently been implemented in the overall AspenPlus process model. In parallel we performed experiments on $CO_2$ removal with the aqueous MDEA-co-solvent on our absorption-regeneration pilot in Lacq, to validate the results of the process simulations. The results of this work will be presented and the overall impact of adding a significant amount of physical co-solvent will be discussed.

# 625635 The Investigation of Impact of Mineralogical Heterogeneity of Clay–Calcite Based Fracture–Matrix System for CO2 Storage By Using a Hybrid–Scale Model

**Abstract:**

CO2 storage and sequestration are considered an effective approach to mitigate greenhouse gas emissions. Injected CO2 in the subsurface formation can easily move upward by buoyancy or through leakage out of the target reservoir through the existing fractures and negatively influences the CO2 sealing capacity and security. While injecting an enormous amount of CO2 into carbonate–rich aquifers, CO2 is effortlessly dissolved in the formation brine under the high pressure, and the subsequently formed CO2–enriched brine can react with the calcite. Reaction–induced changes in pore structure and fracture geometry alter the porosity and permeability, giving rise to the concerns of CO2 storage capacity and security. Previous experimental investigation of a limestone covered with clay coating showed that the clay coating was progressively developed as calcite was dissolved. The dissolution rate of calcite was significantly decreased. The clay swelling decreased the permeability and enhanced the CO2 storage safety capacity. This study aims to analyze the acid–rock interaction, subsequent flow path evolution and ions distribution in the calcite–clay matrix systems during CO2 injection by implementing a multi–scale reactive transport modeling. The dynamic change of calcium ion was tracked because the produced calcium ion concentration decreased the local dissolution rate of calcite dissolution. The injected water mixed with calcium ions was unstable and resulted in the precipitation of CaSO4, causing scale formation. During the injection of CO2–enriched brine, the clay was regarded as nonreactive and considered its swelling effects. Other minerals such as quartz were regarded as nonreactive, while the calcite was reacted with carbonic acid. We elucidated the heterogeneity of mineralogical distribution on the transport pattern and mineral dissolution–induced porosity–permeability relationship. We considered the clay phase of mineralogy in CO2 transport and storage, which gives rise to a more accurate description of the underground geology process and prediction of long–term CO2 exhibition. We implemented immersed boundaries condition (IBC) to describe fluid zones, chemically reacted–phase zone, nonreacted–phase zone, and the plastic–solid swelling phase zone with irregular configurations by the structured grids. The IBC method took advantage of the uniform, structured grids while making it possible to solve the partial differential equations robustly and efficiently. Our simulation results can help establish and calibrate the Darcy–scale or field–scale model to track the CO2 trajectory, design, and optimum CO2 storage project.

To capture this complex sub–surface process, we coupled Darcy–Brinkmann–Stokes (DBS) method with geomechanics of the microporous plastic solid. The calcite phase was regarded as pore–scale in our model, where the clay phase was regarded as Darcy–scale. The DBS momentum equation was used to calculate the fluid flow in the porous media, and a momentum conservation equation for a plastic solid was used to describe clay swelling and deformation. Reactive transport was described by the multi–component advection–diffusion equation. The mass balance equations were used to update the volume fraction of the calcite phase and clay phase. The swelling pressure was obtained by the semiempirical formulation, which was determined by clay volume fraction and sodium concentration in the system. The solid effective viscosity was calculated by Herschel–Bulkley non–Newtonian plastic viscosity models. A series of partial differential equations were discretized by applying the Finite Volume Method (FVM). The multi–scale multi–species reactive transport model was developed based on the pisoFoam framework in OpenFOAM, an incompressible transient flow solver. The discretized algebraic equations were solved subsequently. To validate the numerical model in simulating the multi–species reactive transport, we compared the numerical results with the analytical solution of the Plane Poiseuille flow and Kinetic Decay–Chain, respectively. The numerical results were matched well with the analytical results, showing the confidence of our simulation model.

We analyzed the heterogeneity of mineralogical distribution in the ideal fracture–matrix models. The mineralogical matrixes were symmetrically distributed at the upper and bottom sides of the fracture. Several distribution patterns with the same porosity were considered in the investigation, such as clay coating, striped distribution, staggered pattern, random distribution, and clustered distribution. From the simulation, we have the following interesting findings. First, under different clay–calcite distributions, the local dissolution rate of calcite and calcium ion accumulation were various. For clay coating pattern, only slight dissolution was observed at inner calcite, because the clay distributed around the calcite swelled and subsequently reduced the permeability. The contacted surface area between CO2–enriched brine and calcite decreased. Due to the fluid flow limitation in the fracture, CO2–enriched brine cannot saturate well the fracture. When the calcite was mainly distributed at the downstream area, there was no apparent dissolution of calcite. However, when the calcite was mainly distributed around the inlet area, the calcite showed significant dissolution. Second, we also found that the clay swelling size had no significant relations with their distribution in the system. The clay distributed in the different areas has the same swelling size. However, under the action of fluid flow, the clay showed the deformation in the system. As clay clustered together, only outside layers of clay have obvious deformation. For the random distribution pattern, calcite was fully mixed with clay. Most clay in the system has deformation. Third, we observed meaningful phenomena regarding calcium ion concentration. Calcium ion concentration can reflect the underground reaction rate of dissolution. Total calcium ion concentration increased initially in the system. As calcite dissolved, calcium concentration decreased, and then kept a constant value. Nevertheless, the mineralogical heterogeneity has significant impacts on the total calcium. For example, the clay–coating pattern accumulated more amount in the system. The inner calcite was dissolved and produced calcium, while as the outer clay swelled, the permeability decreased, and significant amount of calcium was accumulated. When calcite was distributed all around the fracture, the calcium ion was not likely to be accumulated in the system. If it was the clay distributed all around the fracture, the generated calcium tended to accumulate around the upper and bottom wall due to the viscosity limitation of the boundaries. The porosity–permeability relationships of different patterns were obtained, and we presented a universal formula to describe the relationship by introducing the geometrical factors of the distribution pattern.

From the research, we found the several mineralogical patterns with better storage capacity, including clay distributions around the calcite, fracture, and upstream side. Clay swelling can significantly decrease the permeability and limited the transport of CO2–enriched brine. However, clay decrease the contacting area between the calcite and acid, preventing the dissolution of carbonate. Understanding the flow pattern and dissolution characteristics can ensure that CO2 has been stored in the target zone and whether CO2–enriched brine leaked through the channels. The pore–scale parameters from this investigation can help establish a Darcy–scale model to track the trajectory of CO2 in the subsurface, which can help enhance the CO2 sealing capacity and guarantee environmental security.

# 629172 Enhancement of Thermal and Hydraulic Characteristics of Microchannel Heat Sinks By Employing $sCO_2$ As Coolant

**Abstract:**

Micro-scale thermal management has become a challenge currently with the extraordinary shrinkage of microchips in computing and communication devices [1]. The operating temperatures associated with microchips must be kept within stipulated limits that required cooling systems that are extremely compact and exhibit excellent heat transfer characteristics. Besides, the heat flux rates from microelectronics have gone out to be considerably above the projected values for the year 2020 [2] with recent nanotechnological advancements. Henceforth, the compact heat removal system's design has become intricate, requiring novel coolants and heat sink geometries to enhance thermal and hydraulic associated with the cooling systems.

Liquid-cooling systems have already replaced the air-cooling systems due to the limited heat removal rates associated with the air, but further research is still needed for superior thermal cooling systems. The microchannel heat sink presents itself as a favorable solution, as demonstrated by its compact size and high heat dissipation capabilities [2]. However, extremely high-pressure drops associated with the microchannel heat sinks (MCHS) are a concern that needs to be addressed to reduce the pump size and the overall size of the heat-removing systems. The literature shows that an appropriate choice of coolant can play a crucial role in improving a microchannel heat sink's performance. Various coolants have been examined for MCHS by different researchers, such as methyl alcohol [1], R410a [2], and R134a [3]. Additionally, nanofluids initially industrialized by Choi and Eastman [4] have recently gained researchers' attention. A recent study [5] indicated that thermohydraulic characteristics associated with the mini channel heat sinks could be enhanced considerably using alumina nanofluid. Similar findings have been reported for the copper (Cu)/water [6], copper-oxide(CuO) /water [7], diamond/glycerin and diamond/water [8], $TiO_2$/water [9], and ZnO/ethylene glycol [10]. But at the same time, studies [6-10] suggest the improved thermal performance utilizing nanofluids comes at the cost of pressure drops that in turn will increase the pump size.

The literature discussed above indicated a dire need for a coolant that can enhance both thermal and hydraulic characteristics of MCHS for an efficient cooling system. In this reference, supercritical carbon dioxide ($sCO_2$) operating close to its critical point can be of great value because of its promising thermophysical properties. Hence, the thermohydraulic performance of the $sCO_2$-cooled microchannel heat sinks (MCHS) are investigated numerically and compared with the conventional water-cooled MCHS in the current work for the first time to the author's best knowledge. Header geometry for both $sCO_2$-cooled and water-cooled MCHs are also optimized for uniform flow distribution; else, maldistribution can considerably influence the performance of the MCHS [11–15]. Finally, the effect of different operating conditions on the thermohydraulic performance of the proposed $sCO_2$-cooled MCHS is examined to locate its optimal operating conditions. To perform the conjugate heat transfer calculations mentioned above, a 3D Reynolds Averaged Navier–Stokes (RANS) model is established and validated for both coolants, and shear stress transport (SST) turbulence model is used to model the flow turbulence. The computational geometry of both fluid and solid domains and its hexahedral mesh generated using ANSYS ICEM-CFD are shown in Fig. 1a. The mesh's same topology was enforced in the interface between fluid and solid domain to minimize the interface losses. Thermophysical properties of the changes substantially near the critical point, and it is essential, therefore, to integrate these variations in ANSYS-CFX during the numerical calculation for accurate prediction of the thermohydraulic characteristic of $sCO_2$-cooled MCHS. Thermophysical properties of are incorporated via a high-resolution real gas property (RGP) file to capture sharp changes in its properties.

Results suggest both distributor and collect header geometry can impact the thermohydraulic performance of the MCHS severely. It is found that rectangular header geometry minimizes the flow maldistribution (Fig. 1b) and results in a uniform temperature of the MCHS's base. The comparison of the performance of the MCHS with two coolants reveals that $sCO_2$-cooled MCHS is up to 32% higher in comparison with water-cooled MCHS (Fig. 1d). Further base temperature with the former was found less in comparison with the latter. It is also found that the $sCO_2$-cooled MCHS can withstand the off-load heating loads without any change in the base temperature. Moreover, it is found that the $sCO_2$-cooled MCHS reduces the pressure drop up to 7 times (Fig. 1e) compared to water-cooled MCHS, while the friction factor for the proposed coolant is up to 10 times smaller than the water-cooled MCHS. Finally, based on overall performance evaluation criteria performance of $sCO_2$-cooled MCHS was found up to 2.2 times better (Fig. 1f) than the water-cooled MCHS.

**References**

[1] X.F. Peng, G.P. Peterson, B.X. Wang, HEAT TRANSFER CHARACTERISTICS OF WATER FLOWING THROUGH MICROCHANNELS, Exp. Heat Transf. 7 (1994) 265–283. https://doi.org/10.1080/08916159408946485.

[2] R. Yun, J. Hyeok Heo, Y. Kim, Evaporative heat transfer and pressure drop of R410A in microchannels, Int. J. Refrig. 29 (2006) 92–100. https://doi.org/10.1016/j.ijrefrig.2005.08.005.

[3] G. Li, X. Fang, Y. Yuan, Y. Chen, L. Wang, Y. Xu, An experimental study of flow boiling frictional pressure drop of R134a in a horizontal 1.002 mm tube under hypergravity, Int. J. Heat Mass Transf. 118 (2018) 247–256. https://doi.org/10.1016/j.ijheatmasstransfer.2017.10.120.

[4] J.A. Eastman, U.S. Choi, S. Li, L.J. Thompson, S. Lee, Enhanced Thermal Conductivity through the Development of Nanofluids, MRS Proc. 457 (1996) 3. https://doi.org/DOI: 10.1557/PROC-457-3.

[5] M. Saeed, M.-H. Kim, Heat transfer enhancement using nanofluids (Al2O3-H2O) in mini-channel heatsinks, Int. J. Heat Mass Transf. 120 (2018) 671–682. https://doi.org/10.1016/J.IJHEATMASSTRANSFER.2017.12.075.

[6] Z. Azizi, A. Alamdari, M.R. Malayeri, Thermal performance and friction factor of a cylindrical microchannel heat sink cooled by Cu-water nanofluid, Appl. Therm. Eng. 99 (2016) 970–978. https://doi.org/10.1016/J.APPLTHERMALENG.2016.01.140.

[7] S.M. Peyghambarzadeh, S.H. Hashemabadi, A.R. Chabi, M. Salimi, Performance of water based CuO and Al 2 O 3 nanofluids in a Cu–Be alloy heat sink with rectangular microchannels, Energy Convers. Manag. J. 86 (2014) 28–38. https://doi.org/10.1016/j.enconman.2014.05.013.

[8] H.A. Mohammed, P. Gunnasegaran, N.H. Shuaib, Influence of various base nanofluids and substrate materials on heat transfer in trapezoidal microchannel heat sinks, Int. Commun. Heat Mass Transf. 38 (2011) 194–201. https://doi.org/10.1016/j.icheatmasstransfer.2010.12.010.

[9] W. Duangthongsuk, S. Wongwises, Heat transfer enhancement and pressure drop characteristics of TiO 2 –water nanofluid in a double-tube counter flow heat exchanger, (2008). https://doi.org/10.1016/j.ijheatmasstransfer.2008.10.023.

[10] J.K. Kuo, C.N. Huang, T.H. Lu, Experimental investigation on heat transfer and pressure drop of ZnO/ethylene glycol-water nanofluids in transition, Appl. Therm. Eng. 93 (2016) 425–432. https://doi.org/10.1016/j.applthermaleng.2015.09.100.

[11] S. Kim, E. Choi, Y.I. Cho, Effect of header shapes on the flow distribution in a manifold for electronic packaging applications, Int. Commun. Heat Mass Transf. 22 (1995) 329–341. https://doi.org/10.1016/0735-1933(95)00024-S.

[12] R. Chein, J. Chen, Numerical study of the inlet/outlet arrangement effect on microchannel heat sink

performance, Int. J. Therm. Sci. 48 (2009) 1627–1638. https://doi.org/10.1016/j.ijthermalsci.2008.12.019.

[13] E.S. Cho, J.W. Choi, J.S. Yoon, M.S. Kim, Experimental study on microchannel heat sinks considering mass flow distribution with non-uniform heat flux conditions, Int. J. Heat Mass Transf. 53 (2010) 2159–2168. https://doi.org/10.1016/j.ijheatmasstransfer.2009.12.026.

[14] E.S. Cho, J.W. Choi, J.S. Yoon, M.S. Kim, Modeling and simulation on the mass flow distribution in microchannel heat sinks with non-uniform heat flux conditions, Int. J. Heat Mass Transf. 53 (2010) 1341–1348. https://doi.org/10.1016/j.ijheatmasstransfer.2009.12.025.

[15] R. Manikanda Kumaran, G. Kumaraguruparan, T. Sornakumar, Experimental and numerical studies of header design and inlet/outlet configurations on flow mal-distribution in parallel microchannels, Appl. Therm. Eng. 58 (2013) 205–216. https://doi.org/10.1016/j.applthermaleng.2013.04.026.

a. Continuation domain and mesh

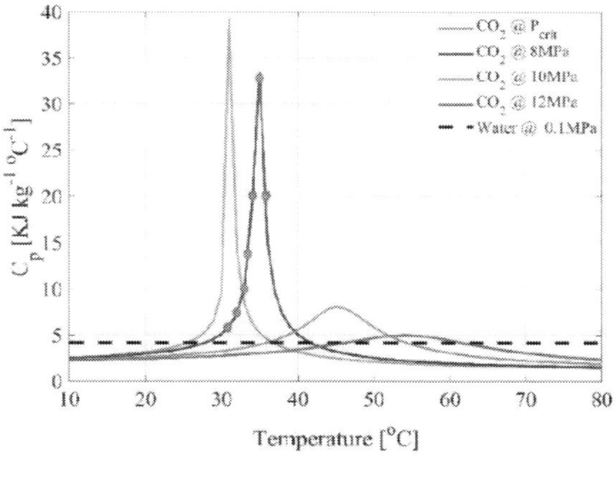


b. Comparison of the Cp of water and $sCO_2$, red dot showing the inlet temperature conditions

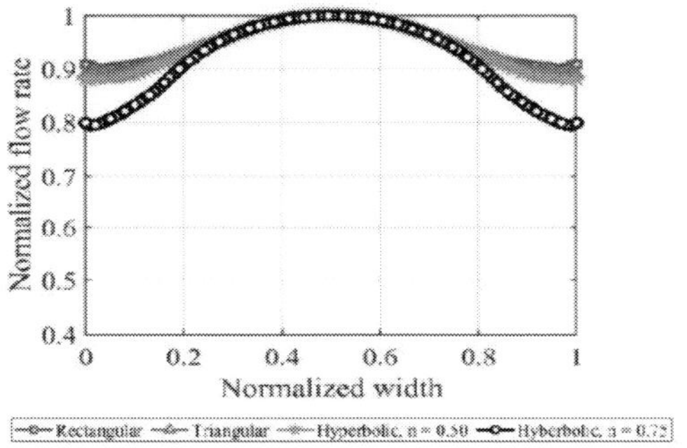


c. Normalized flow rates along the width of the heat sink

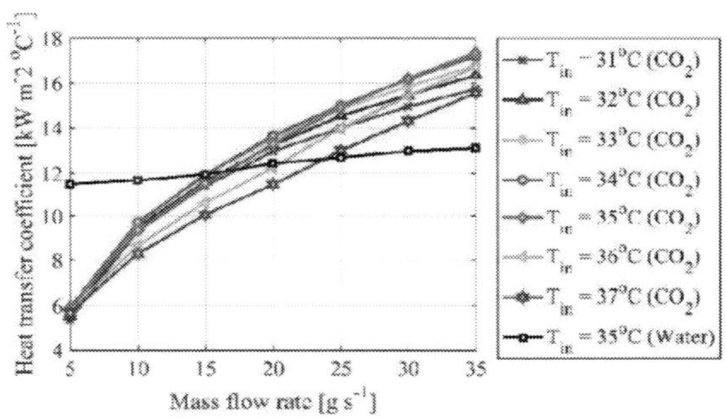


d. Heat transfer coefficient comparison of sCO_2-cooled MCHS and water-cooled MCHS

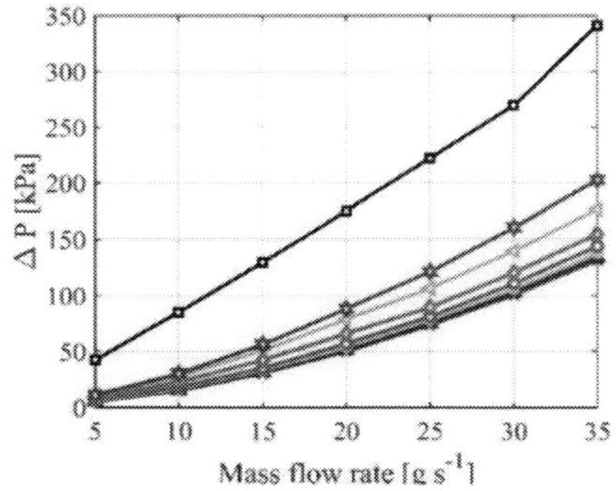


e. Preressur drop comparisson of $sCO_2$-cooled MCHS and water-cooled MCHS

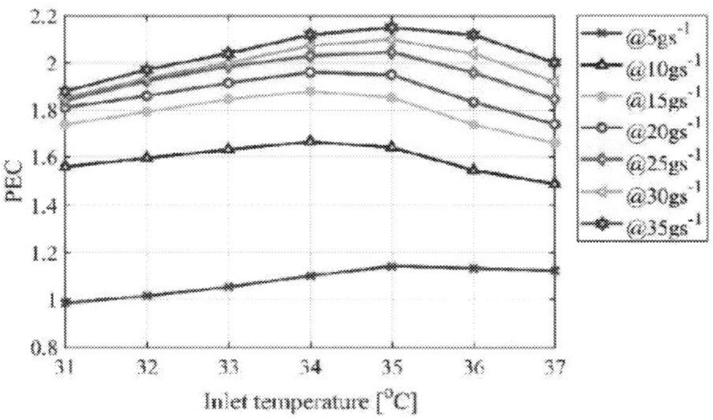


f. Performance evaluation criteria of $sCO_2$-cooled MCHS at different flow rates and inlet temperature condition of the coolant

# 627135 CFD-Based Modelling of an Open Flow Field Architecture Proton Exchange Membrane Fuel Cells

**Abstract:**

Nuvera fuel cell used an advanced technology to design fuel cells suited to the demanding needs of heavy-duty vehicle application. This technology incorporates an open flow field architecture that enables higher power density and increases the active area of each cell MEA. In this work, a 3D computational fluid dynamic (CFD) model for proton exchange membrane (PEM) fuel cells has been developed. The model accounts for convective and diffusive transport of species in the open flow fields, GDL, catalyst and membrane layers with electrochemical reactions. The heat generation was calculated with the electrochemical reaction, ohmic resistance and phase change due to condensation/vaporization of water. The model solves for the electric and ionic potentials in the electrodes and membrane. This model predicts the local current density, overpotential, and water content in all of the layers, including the open flow field cathode and anode. The model showed a good agreement to the experimental polarization curve, cathode pressure drop and coolant outlet temperature. The model suggested the local flow and temperature distributions have strong effect on the current density distribution in the open field fuel cell. The model was able to predict the fuel cell performance at the start up transient condition.

# 630603 In-Line Analysis of Fuel Cell Materials – Detection and Impact of Defects in Membrane and Catalyst

**Abstract:**

Fuel cells are actively being commercialized for large-scale applications, such as zero-pollution automated systems, but are held back by high manufacturing costs, expensive catalysts, and the fuel delivery cost. While significant cost reductions have been made, the membrane electrode assembly components alone account for as much as 43% of the total system cost of a 2020 auto system at 1,000 units/year and 23% at 500,000 units/year. Fundamental improvements to membrane electrode assemblies (MEA) are occurring rapidly in many ways, such as through reduced catalyst loading, alternative catalysts, novel electrodes, improved electrode fabrication, and better assembly processes. However, quantifying improvements can be challenging due to small-scale assembly practices that induce non-uniformities, or defects, into an MEA. Current manufacturing practices rely heavily on cell testing to indicate material issues, which leads to delayed correction of production issues. At the research scale, all variation needs to be controlled so the changing variables (e.g., loading) have their effects properly characterized. Mainstream Engineering has developed and evaluated an in-line optical scanner that can determine membrane thickness, identify defects, and measure catalyst loading and variability. The inspection system samples every location so that defects in the materials can be selectively removed prior to assembly. This technology can improve the repeatability of data on new technologies and enhance rapid prototyping.

A variety of materials have been examined including proton exchange membrane (17 – 250 µm thick, supported and unsupported), anion exchange membrane (30 – 50 µm thick, reinforced), hydrocarbon membrane (20 – 165 µm thick, unsupported), catalyst coated membrane (Pt and Ir, 0.1 – 0.4 $mg/cm^2$), gas diffusion electrodes (Pt/C 0.1 – 0.4 $Pt/cm^2$), and a variety of support/protective layers. Defects in these materials (from 5 – 500 µm in size) were induced and optical features determined. The impact of induced pinhole defects on MEA performance and lifetime were examined through initial testing and accelerated stress tests. These tests provide a baseline for what defects must be seen, where they impact performance, and how improved in-line quality control can reduce assembled component failures. Sensor development, capabilities, impact of defects on membrane and fuel cell performance, and cost benefits during manufacturing and testing will be presented.

# 630814 Highly Dispersion Nickel-Molybdate Bi-Metallic Nanoparticle on Cerium-Zirconium-Yttrium Support for SOFC Application

**Abstract:**

**Highly Dispersion Nickel-Molybdate Bi-metallic Nanoparticle on Cerium-Zirconium-Yttrium support for SOFC application**

***Mohamed A. Elharati***[a], Martinus Dewa[b], Steven R. Saunders[a], M. Grant Norton[a,b], Yohei Miura[c], Song Dong[c], Yosuke Fukuyama[c], Nilesh Dale[d], A. Mohammed Hussain[d,*], Su Ha[a,*]

[a]Voiland School of Chemical Engineering and Bioengineering, Washington State University, Pullman, WA 99164-2710, USA

[b]School of Mechanical and Materials Engineering, Washington State University, P.O. Box 642920, Pullman, WA 99164-2920, USA

[c]Nissan Research Centre, Nissan Motor Corporation Limited, Kanagawa 237-8523, Japan

[d]Nissan Technical Centre North America, Farmington Hills, MI-48335, USA

## Abstract

The ethanol steam reforming (ESR) is a promising alternative energy production technique from renewable resources for Solid Oxide Fuel Cell (SOFC). Herein, the bimetallic supported NiMo/CZY catalyst demonstrates high catalytic activity and stability toward ethanol steam reforming (ESR) reaction and excellent coking resistance than mono-metallic Ni catalyst without Mo. The catalytic activity of synthesized Ni/CZY with and without Mo was conducted at 800-400 °C and an S/C ratio of 2, with a weight-hourly-space-velocity (WHSV) of 156 $h^{-1}$. The XRD analysis illustrates that the bimetallic NiMo/CZY catalyst has high resistance to coking and sintering compared to mono-metallic Ni/CZY catalyst. The $H_2$-TPR showed that NiMo/CZY has a lower reduction temperature than Ni/CZY due to NiMo and CZY support's high interaction. This catalytic enhancement can be associated with the high dispersion of NiMo nanoparticles over CZY support material and the high oxygen storage capacity of CZY support material. This unique material, NiMo/CZY catalyst, was introduced as a micro-reforming active layer on the Ni-YSZ anode supported single SOFC cells. The SOFC cells with a coated layer of and infiltrated NiMo/CZY illustrated high electrochemical performance and stability with a low degradation rate for ~ 100 hours.

**Keywords:**

Ethanol steam reforming, hydrogen production, Ni-Mo nanoparticles catalysts, CZY oxide support material, SOFC

# 630956 Metal-Supported Solid Oxide Fuel Cell System with Infiltrated Reforming Catalyst Layer Operating Under Direct Ethanol Feed Configuration

**Abstract:**

Metal supported solid oxide fuel cell (MS-SOFC) has an advantage over the conventional cermet-based solid oxide fuel cell (SOFC) because MS-SOFC can withstand rapid start-up and cool-down without cracking, making it suitable for mobile application. Meanwhile, the cermet-based SOFC is prone to crack from thermal shock due to the low toughness of its ceramic components. Applying a micro-reforming catalyst layer on the anode surface is an effective solution to prevent cell deactivation from coking formation if hydrocarbon or logistic fuels is used to run the SOFC. However, using a separate reforming reactor or adding a physical catalyst layer over the anode will increase the overall system's volume. Since the anode material is typically introduced on the MS-SOFC using a precursor infiltration method, introducing the reforming catalyst using the same method will keep the system compact.

In this work, we integrate the fuel cell system and reforming catalyst by infiltrating a 5 wt.% Rh/CZ catalyst as an internal reforming catalyst layer on a button MS-SOFC that will run under ethanol steam reforming condition. The catalyst was applied by infiltrating Rh, Ce, and ZrO precursors in the SOFC's metal support. The reforming catalyst will convert the aqueous ethanol solution into the hydrogen, and CO gas stream as an internal catalytic reforming layer is tested under direct-fed ethanol condition (S/C ratio = 2, 600 °C) for 130 hours. The button cell with the infiltrated 5 wt.% Rh/CZ shows improved maximum current density and excellent stability over the button cell without the catalyst layer. Post-test sample analysis reveals that the cell with the infiltrated catalyst can protect the anode functional layer from coking deposition. The proposed integrated reforming catalyst and MS-SOFC system can be a viable solution for a future electric vehicle with bioethanol fed-SOFC technology.

# 626701 Enhanced Durability of $Li_{1.2}Mn_{0.54}Co_{0.13}Ni_{0.13}O_2$ Cathode Material By Slightly Fluorinated $Al_2O_3$ ALD Coating

**Abstract:**

Li-rich layered oxides have been in focus because of their high specific capacity. However, they still have severe challenges, such as phase transition, mechanical cracking, electrolyte decomposition, transition metal dissolution, poor rate performance, and instability in air. To address the interfacial challenges and achieve a stable cathode-electrolyte interphase, surface modification of cathode materials is the most effective approach to prevent electrolyte decomposition and transition metal dissolution. The basic requirements of the coating layers are (1) controllable thickness, (2) uniform and conformable coating on cathode material surfaces, and (3) good mechanical properties. In this study, $Li_{1.2}Mn_{0.54}Co_{0.13}Ni_{0.13}O_2$ (NMC) particles were coated with slightly fluorinated $Al_2O_3$ film by atomic layer deposition (ALD) in a fluidized bed reactor. The charge/discharge cycling results of the assembled battery showed that its electrochemical performance was significantly improved, compared to that of the uncoated NMC. In addition, after being stored in air for one year, there was no degradation in the electrochemical performance of the fluorinated $Al_2O_3$ coated NMC, while that of pure $Al_2O_3$ coated NMC showed a significant decrease. Furthermore, full cells based on coated NMC were assembled; the performance of fluorinated $Al_2O_3$ coated NMC showed significant improvement, compared to those assembled from uncoated NMC.

# 627601 Mass Transport Optimization in Direct Formic Acid Fuel Cell Catalyst Layer Via Pore-Former Templating

**Abstract:**

To meet the ever-rising demand for portable power sources, batteries have become a vital part in mobile electronics such as cell phones and laptops. However, the efficiency of batteries is compromised due to exponential degradation over time, lengthy recharging times, and limited charge capacity. Direct formic acid fuel cells (DFAFC) present an alternative to batteries, due to their near instantaneous re-fueling times, higher efficiency, and 24/7 operation capabilities. The source of fuel (formic acid) is sustainable via renewable sources. [1] However, two-phase flow in conjunction with the small pore size between anode catalyst agglomerates (~20 nm) restricts the cell's efficiency. To improve the mass transport liquid reactants (formic acid) and gaseous products (carbon dioxide), a pore-former is integrated into the anode catalyst layer during fabrication and subsequently removed. The additional templated porosity enhances the two-phase mass transport in and out of the anode catalyst layer.

In 2012, pore-former (lithium carbonate, $LiCO_3$) was incorporated and removed from the anode catalyst layer forming ~10 μm pores. [2] Peaking at 17.5 wt% pore-former addition, the increase in porosity resulted in a reduction of the formic acid electrooxidation charge transfer. The increased porosity the disconnection between catalyst agglomerated and reduced the connected electrochemical surface area.

Previous work has been done with the addition of a smaller pore-former, MgO (~50 nm). [3] The intent of the smaller pore-former is to increase the porosity of the anode catalyst layer while retaining the connectivity of the agglomerates. The addition of 25 wt% pore-former increased the electrochemical surface area and cell performance by 293% and 86%, respectfully, compared to an anode catalyst layer with no pore-former. The present work aims study more variations in the wt% of pore-former (0-30 wt%), the effect of pore-former wt% on catalyst ink viscosity, and the effect of catalyst ink sonication with pore-former.

1. Ma, Z., Legrand, U., Pahija, E., Tavares, R.J., and Boffito, C.D., From $CO_2$ to Formic Acid Fuel Cells. **2021**, *60(2)*, 803-815.
2. Bauskar, A.S. and Rice, C.A., Impact of Anode Catalyst Layer Porosity on the Performance of a Direct Formic Acid Fuel Cell. **2012**, *62*, 36-41.
3. Lam, S., Bixby, M.M., and Rice, C.A., Optimization of Mass Transport within Direct Formic Acid Fuel Cell Catalyst Layer via Pore Formers. **2020**, *98*, 355.

# 632973 Fabrication and Testing of Asymmetric Supercapacitors Fabricated with (Mn,Ti)-Oxides and Chtl Derived Hydrochar/Graphene

**Abstract:**

Asymmetric supercapacitors generally possess higher energy density as well as power density. Supercapacitors of this kind utilize dissimilar electrodes. As graphene is very expensive electrode material, we derived hydrochar electrode from the catalytic hydrothermal liquefaction (CHTL) processing of biomass waste residue such as corn stover. Hydrochar was used as a opposite electrode with recently reported (Mn, Ti)-mixed perovskite oxide electrode and specific capacitance was tested. To prepare hydrochar with superior charge storage capability, the CHTL derived char was thermally/chemically post-treated and used as electrode material for asymmetric supercapacitors. . As-fabricated asymmetric supercapacitors were charged with 2.0 V and 0.01 A for 10 minutes and their several charging-discharging cycles were analyzed by cyclic voltammetry to determine specific capacitance as well as the peaks corresponding to oxidation and reduction potentials. Voltammograms of the asymmetric supercapacitors obtained with hydrochar with (Mn,Ti)-oxide opposite electrodes were compared with graphene/(Mn,Ti)-oxide electrodes. Results obtained with specific capacitance will be presented.

# 630236 Redox Thermodynamics of Fe-Doped Magnesium Manganate for Thermochemical Energy Storage

**Abstract:**

Thermochemical energy storage (TCES) is an alternative technology to sensible and latent heat storage that can provide higher energy storage density. TCES can be paired with existing technologies such as concentrates solar thermal plants or grid electricity to provide long duration energy storage. High-temperature TCES is of particular interest as it can be coupled with high efficiency power cycles providing a unique opportunity for large scale energy storage. Existing research has investigated TCES systems utilizing various storage materials including metallic hydrides, carbonates, hydroxides, and metal oxides. Of these materials, the $Mn_3O_4$/MnO system has been studied for its relatively high reaction enthalpy in addition to a low cost, however, it faces limitations as it is susceptible to sintering reducing its reversibility and energy storage density upon cycling. Combining magnesium with manganese oxide in $MgMn_2O_4$ actively reduces the sintering of manganese, providing full reversibility over hundreds of cycles as well as a large energy storage density when combining sensible and thermochemical heat. Further improvements are envisioned by the addition of a small concentration of dopants to increase energy storage density by changing the underlying thermodynamics of the material. In this work, we will present how doping $MgMn_2O_4$ with Fe allows for increased energy density while maintaining reactive stability. The fundamental thermodynamic description of the materials reduction and oxidation behavior derived from van't Hoff analysis using thermogravimetric data will be presented. The material characterization confirms the phase purity and phase transitions of the doped material.

# 631948 Electrochemical Intercalation of Protons into Fluorine-Doped Molybdenum Oxide: A Positive Electrode for Rechargeable Aqueous Aluminum Metal Batteries

**Abstract:**

The global demand for low cost, energy dense electrochemical energy storage technologies has increased significantly in recent years due to the need to integrate renewable energy sources into the electrical grid and electrify transportation systems. Researchers have been particularly interested in batteries using aqueous electrolytes because of their low cost, safety, environmental friendliness, and ability to scale. However, one key issue with aqueous batteries is that their specific energy densities are lower compared to state-of-the-art non-aqueous systems, a consequence of the low electrochemical stability window of water (1.23 V). Strategies to overcome this challenge include "water in salt electrolytes" (WiSE), which permit higher cell voltages, though high electrolyte costs for WiSE-based batteries remain a challenge. An alternative approach to enhancing energy density while maintaining the advantages of traditional aqueous electrolytes is to pair the aqueous electrolyte with a high-capacity metal anode whose corresponding metal cation is multivalent, enabling multiple electron transfers per electroactive ion. Yet other than zinc, the majority of earth abundant multivalent metal anodes are not electrochemically compatible with water in rechargeable battery systems. Recently, it has been shown that artificial interfaces on aluminum metal anodes can enable reversible cycling in aqueous electrolytes, though much remains to be understood regarding the reversible electrodeposition of aluminum metal in such systems. In addition, few positive electrodes have been successfully paired with them to date, a consequence of the challenges associated with electrochemically intercalating highly charged $Al^{3+}$ cations into electrode host structures.

Here, we report fluorine (F)-doped $\alpha$-$MoO_3$ nanowires as a high-capacity positive electrode material for rechargeable aqueous aluminum batteries and study its ion charge storage mechanism. Using a combination of electrochemical measurements, solid-state nuclear magnetic resonance (NMR) spectroscopy, and X-ray diffraction (XRD), we investigate the nature of the electroactive ion, crystal structure changes upon ion (de)intercalation, and electrochemical effects of F-doping the $\alpha$-$MoO_3$ electrodes. A high discharge capacity of 350 mAh/g was achieved upon first discharge, which faded upon continued cycling due to a combination of both cathode and anode effects. Solid-state $^{1}H$, $^{19}F$, and $^{27}Al$ single-pulse and -dipolar-mediated NMR experiments performed at different states-of-charge establish proton intercalation upon discharge, while possible co-intercalation of hydrated aluminum cations was investigated. Quantitative solid-state $^{1}H$ NMR measurements were correlated with Coulomb-counting electrochemical measurements to yield further insights into the ion intercalation processes. F-doping is also shown to enhance solid-state ion transport, rate performance, and cyclability of the $\alpha$-$MoO_3$ electrodes. Overall, the results establish F-doped $\alpha$-$MoO_3$ as a promising proton intercalation electrode material for aqueous batteries, highlight the role of proton intercalation in aqueous electrolytes containing metal salts, and help lay the scientific groundwork towards rechargeable aqueous aluminum metal batteries.

# 628628 Design and Characterization of a Neutron-Friendly Lithium-Ion Battery Coin Cell for Extreme Fast-Charging

**Abstract:**

Lithium-ion batteries (LIBs) have profoundly advanced the development of electric vehicles (EV). However, one of the remaining bottlenecks in the widespread deployment of EVs is the long charging time typically required for commercial LIBs. There is a global push towards extreme fast charging (XFC) of EV batteries to reduce their charging times to 10-15 minutes.[1] However, XFC causes severe degradation in the electrochemical performance of LIBs, which is mainly attributed to irreversible Li plating on the surface of the graphite anode rather than intercalation of Li ions inside the graphite as the battery is fast-charged.[2-3] Irreversible Li plating during XFC has been characterized using a variety of techniques such as scanning electron microscopy, transmission electron microscopy, X-ray diffraction, and X-ray micro-computed tomography (CT) amongst others.[4] While valuable, these characterization techniques have limitations in terms of the sample preparation involved, spatial and temporal resolution required, and/or the necessary imaging contrast to distinguish between materials of similar atomic number (Z). In addition, understanding the origin and spatial heterogeneity of Li plating across three dimensions (3D) on porous graphite anodes as a function of time necessitates a non-destructive characterization platform to differentiate between graphite and Li.

Neutron-based imaging provides the sensitivity to differentiate graphite from Li due to the large difference in their total neutron (absorption, coherent, and incoherent scattering) cross-sections (Li:C ~ 73:11 barns)[5] that are challenging to detect using X-ray imaging alone. However, high energy X-rays provide the benefit of penetrating the metallic battery components. e.g., Cu, and Al current collectors. Therefore, to test the feasibility of simultaneous neutron and X-ray tomography (NeXT), we performed proof-of-concept multi-modal imaging experiments at the BT-2 imaging beamline at National Institute of Standards and Technology Center for Neutron Research.[6-7] Our spatial resolution for both X-rays and neutrons was ~15-20 μm. For ex-situ Li plating detection, we imaged 1) pristine/uncycled and 2) a cycled graphite anode. For cycled graphite anode, we disassembled the battery pouch cells in their fully discharged condition and harvested the anode after 450 cycles of 9C charging and C/2 discharging. Thus, our cycled anodes contained regions of high and low Li plating. In addition, to determine NeXT feasibility for in-situ imaging, we characterized a fully assembled, uncycled single-layer battery pouch cell.

For data analysis, we used 2D bivariate histogram phase segmentation to differentiate different battery materials such as Cu, graphite and Li, Al, etc.[8] However, too low of an X-ray energy (40keV) combined with the flat sample geometry of the graphite strips made it challenging to remove beam hardening artifacts from copper and, therefore, to segment lithium from graphite. For the fully assembled, uncycled pouch cell, we gently rolled the pouch into a ring to fit it into the field-of-view for the highest resolution settings of the system. However, the high neutron attenuation by the hydrogen in EC: EMC (3:7) electrolyte and hydrogenous polymeric materials in the Celgard separator made it particularly difficult to segment graphite anode from the separator and positively identify interfaces.

In this presentation, we will outline the design and characterization of a neutron-friendly LIB coin cell for extreme fast-charging. Our design is based on a standard 2032 format coin cell with custom modifications to address the dual criteria of electrochemistry and neutron imaging, i.e., good neutron transmission and relative contrast from a functional LIB at 6-C charging rate. We will present three advantages of coin cell geometry: 1) coin cells provide more uniform attenuation throughout the entire tomography scan thus reducing artifacts; 2) multiple coin cells can be stacked on top of each other for high throughput battery imaging experiments; 3) they reduce the amount of expensive deuterated materials needed while developing the optimum cell configurations. In addition, we will address the need of a deuterated EC: DMC electrolyte in a fluorinated separator for the neutron-friendly battery design (Figure 1). Since the total neutron cross-section of deuterium and fluorine are on the orders of magnitude lower than that of hydrogen, we anticipate that deuterated electrolyte and a fluorinated separator will help on several fronts. First, these modifications will increase neutron transmission through the battery, thus improving signal-to-noise ratio and reducing acquisition time especially at low neutron flux facilities. Secondly, they will result in a cleaner electrolyte-separator-anode interface for the detection of plated lithium because of absence of hydrogen which is a major source of incoherent scattering in the neutron histogram. Finally, we will show how these modifications affect the electrochemical performance of our LIB coin cells to determine the onset of Li plating at 6-C charging rate. i-e., capacity fade vs. number of cycles.

**References**:


1. Liu, Y., Zhu, Y., & Cui, Y. (2019). Challenges and opportunities towards fast-charging battery materials. Nature Energy, 1.
2. Tomaszewska, A., Chu, Z., Feng, X., O'Kane, S., Liu, X., Chen, J., ... & Li, Y. (2019). Lithium-ion battery fast charging: A review. eTransportation, 1, 100011.
3. Liu, Q., Du, C., Shen, B., Zuo, P., Cheng, X., Ma, Y., ... & Gao, Y. (2016). Understanding undesirable anode lithium plating issues in lithium-ion batteries. RSC advances, 6(91), 88683-88700.
4. Paul, P. P., McShane, E. J., Colclasure, A. M., Balsara, N., Brown, D. E., Cao, C., ... & Nelson Weker, J. (2021). A Review of Existing and Emerging Methods for Lithium Detection and Characterization in Li-Ion and Li-Metal Batteries. *Advanced Energy Materials*, 2100372.
5. https://www.nist.gov/ncnr/neutron-scattering-lengths-list
6. LaManna, Jacob M., et al. "Neutron and X-ray Tomography (NeXT) system for simultaneous, dual modality tomography." *Review of Scientific Instruments*11 (2017): 113702.
7. Yusuf, Maha, et al. "Lithium Plating Detection in Extremely Fast-Charged Lithium-Ion Batteries Using Simultaneous Neutron and X-Ray Imaging." *ECS Meeting Abstracts*. No. 3. IOP Publishing, 2020.
8. LaManna, J. M., et al. "NIST NeXT: a system for truly simultaneous neutron and x-ray tomography." *Hard X-Ray, Gamma-Ray, and Neutron Detector Physics XXII*. Vol. 11494. International Society for Optics and Photonics, 2020.


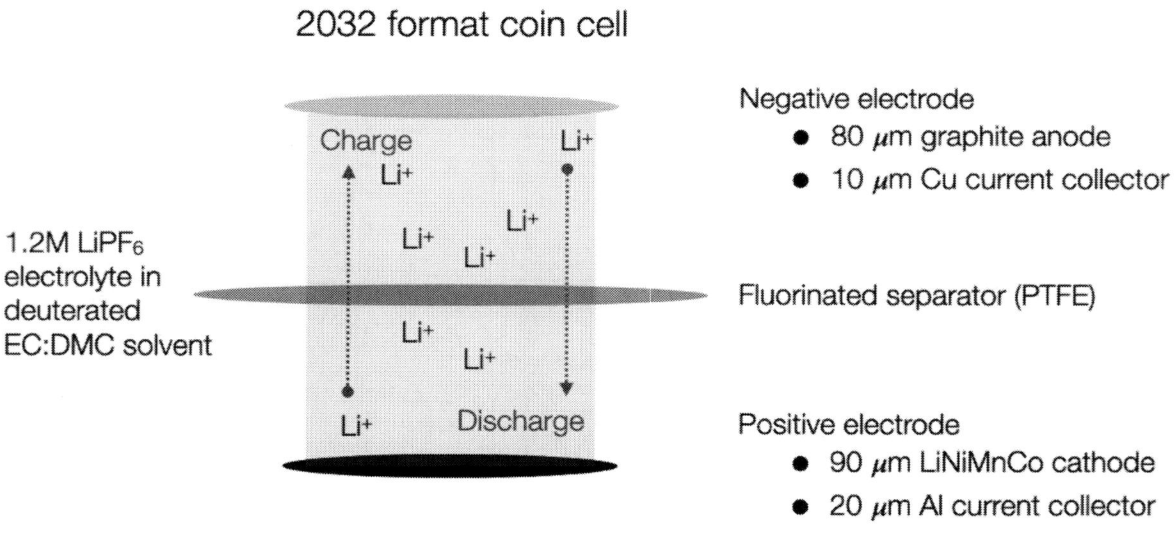


***Figure 1.*** *Schematic of a modified 2032 format coin cell showing battery electrodes and fluorinated separator in a deuterated electrolyte.*

# 628654 In Operando Study of All-Solid-State Lithium Batteries Coupling Thioantimonate Superionic Conductors with Metal Sulfide

**Abstract:**

All Solid-state Lithium metal batteries (ASLBs) are considered as next generation energy storage devices to deliver high energy density and safety. The intrinsic nonflammability of solid-state electrolytes (SEs) can significantly improve the battery safety by reducing fire risk. Among various SEs candidates, sulfide SEs have ultrahigh ionic conductivity and low mechanical stiffness. However, sulfide SEs suffer from poor compatibility with the conventional transition metal oxide electrodes due to narrow electrochemical stability window and poor chemical stability. Therefore, it is crucial to pair sulfide electrolytes with the proper cathode in developing stable ASLBs.

In this work, we synthesized and systematically studied one type of thioantimonate ion conductor, $Li_{6+x}Ge_xSb_{1-x}S_5I$, with different compositions, where $Li_{6.6}Ge_{0.6}Sb_{0.4}S_5I$ exhibits highest ionic conductivity of 1.6 mS $cm^{-1}$ at room temperature. A metal sulfide $FeS_2$ cathode which owns high specific capacity and excellent chemical compatibility with sulfide electrolytes was chosen to couple with $Li_{6.6}Ge_{0.6}Sb_{0.4}S_5I$ in ASLBs without additional interface engineering. For the first time, *in situ* energy dispersive X-ray diffraction (EDXRD) was applied in the study of ASLBs without disassembling the sealed cells or special cell designing to characterize the structural stability of the $Li_{6.6}Ge_{0.6}Sb_{0.4}S_5I$ and $FeS_2$ during cycling. EDXRD in situ observed the decomposition of $Li_{6.6}Ge_{0.6}Sb_{0.4}S_5I$ at low voltage which confirms the safe operation voltage range. The electrochemical stability was investigated, and a safe operating voltage window ranging from 0.7~2.4 V (vs. In-Li) was achieved. Compared with fast-decaying performance in liquid electrolyte batteries, the SEs can effectively confine $Fe^0$ aggregation and polysulfide shuttle effect in the ASLBs. The ASLBs paring $FeS_2$ cathode and $Li_{6.6}Ge_{0.6}Sb_{0.4}S_5I$ electrolyte delivers a high initial capacity of 715 mAh $g^{-1}$ at C/10 which keeps stable for 220 cycles with a high-capacity retention of 84.5%. As a result of excellent compatibility between FeS2 and $Li_{6.6}Ge_{0.6}Sb_{0.4}S_5I$, outstanding capacity and cycling stability are achieved at the same time.

# 637769 Role of physics based models for digital twins

**Abstract:**

Models are imaginary representations of the world that we build to understand it. They help us navigate the complexity involved in the naturally occurring physico-chemical processes. This can be utilized to design engineering solutions that are necessary for as a civilization such as power generation, chemical manufacture and others.

Models can be created based on recorded data (using ML and data analytics) as well as starting from fundamental principles of physics. Building physics based models that capture details and then translate them in to reduce models (rather than other way around) that can be utilized during operations is a prudent way to ensure safety and avoid errors.
The advancement and consistent adoption of cutting-edge digital simulation is critical to this future, and necessary in order to respond to the concerns of the market and of the public. Predictive modeling and simulation (M&S) techniques for reactor thermal hydraulics (TH) are essential to bringing the desired improvements to the economics and safety standards of process & power plants.

In this workshop, starting from the high level concept of digital twins for chemical and nuclear operations, we will dive in to the details of design and eventually capturing flow and turbulence behavior. This will set the importance of accurate physics based models for creating such digital twins. IT will be explained with the help of examples. We will also talk about the tools necessary to accomplish this as well as how they need to be integrated in to a common engineering process.

# 622352 Understanding $V^{2+}/V^{3+}$ Reaction on Metal Electrocatalysts for Vanadium Redox Flow Batteries

**Abstract:**

The world's energy demand, which is expected to grow by 50% by 2050,[1] is currently largely met by burning fossil fuels that emit carbon dioxide (~35 billion metric tons of $CO_2$ released in 2020).[1] The large increase in $CO_2$ in the atmosphere has led to a rise in the Earth's temperature by ~0.40 °C in the past two decades,[2] causing harmful climate change and a huge loss to biodiversity. This rising energy demand can be met sustainably by using energy from clean renewables like solar and wind, which is expected to supply ~56% of the global electricity demand by 2050.[3] However, the intermittent nature of solar and wind necessitates the development of low-cost energy storage technologies such as redox flow batteries (RFBs) for ensuring grid reliability. RFBs store energy in flowing electrolytes rather than in an electrode and are highly promising to store renewable electricity at a large scale because of high power densities, long electrode lifetimes, and their ability to decouple energy and power, enabling high scalability.

All-vanadium redox flow batteries (VRFBs) are the most developed RFBs, but they suffer from high costs partly due to the slow reaction rate at the negative electrode, preventing them from reaching the Department of Energy's capital cost target.[4] VRFBs store energy in different oxidation states of vanadium ($VO_2^+/VO^{2+}//V^{2+}/V^{3+}$) dissolved in acidic electrolytes. $V^{2+}/V^{3+}$ reaction contributes ~80% of the total overvoltages in VRFBs, lowering their energy efficiency.[5] Using active electrocatalysts to eliminate the kinetic overvoltage of $V^{2+}/V^{3+}$ reaction alone will increase the energy storage efficiency from 77 to ~86%, lowering the capital costs.[6]

Certain metal electrocatalysts like Bi,[7] Cu,[8] Sn,[9] and Sb supported on carbon have shown improved $V^{2+}/V^{3+}$ kinetics,[10] but the cause of enhancement is unknown, preventing rational electrocatalyst design. Further, it is uncertain whether the improved kinetics is due to the metal or a change in functional groups (or increase in surface area) on carbon as a result of metal deposition. Our prior work has shown that $V^{2+}/V^{3+}$ reaction proceeds through an adsorbed vanadium intermediate on glassy carbon, and by controlling the energetics of the intermediate, the $V^{2+}/V^{3+}$ charge transfer can be enhanced.[6,11] All these tested metal electrocatalysts adsorb hydrogen weakly, so we hypothesize that weakly adsorbed hydrogen facilitates vanadium intermediate formation by not competing for active sites. If this is true, the hydrogen adsorption ($\Delta G_H$) and vanadium intermediate adsorption ($\Delta G_I$) free energy would control the reaction rate, and there will be an ideal $\Delta G_H$ and $\Delta G_I$ resulting in the highest catalytic activity, based on the Sabatier principle of catalysis. The relation between the electrocatalyst's $\Delta G_H$, $\Delta G_I$, and $V^{2+}/V^{3+}$ reaction has not been previously explored.

In this work, we explore the hypothesis that weakly adsorbed hydrogen and vanadium intermediate leads to higher $V^{2+}/V^{3+}$ activity by conducting kinetic measurements on metal electrocatalysts (unsupported on carbon) with varied $\Delta G_H$ and $\Delta G_I$. We extract exchange current densities ($i_o$), apparent frequency factors, and apparent activation energies ($E_a$) from experimental steady state current and electrochemical impedance spectroscopy measurements conducted at various rotation rates, using a rotating disk electrode setup. The measurements at various rotation rates allows us to deconvolute the kinetic and mass transfer contributions, and hence obtain an accurate estimation of the kinetic parameters. We test metals with varying $\Delta G_I$ evaluated by density functional theory and that adsorb hydrogen strongly (W), weakly but experimentally untested (Au, Ag), and weakly but previously shown to improve $V^{2+}/V^{3+}$ performance (Bi, Cu) (**Figure 1a**). The electrochemical active surface area (ECSA) of these electrocatalysts are measured through either underpotential deposition (Cu, Pb), or $N_2O$ titration method depending on the metal electrocatalyst. The ECSA normalized rates allows fair comparison between activities of various electrocatalysts, which we use to infer mechanistic insights of $V^{2+}/V^{3+}$ reaction on metal electrocatalysts.

We show that the metal electrocatalysts are ~100 times more active compared to glassy carbon for $V^{2+}/V^{3+}$ reaction (**Figure 1b**), and their activity trend is correlated with the adsorption energy of vanadium intermediate. The $i_o$ and $E_a$ on these metal electrocatalysts varies with State of Charge (SoC = $[V^{2+}]/([V^{2+}] + [V^{3+}])$) similar to glassy carbon, confirming the presence of an adsorbed intermediate. The trend in the apparent frequency factors allow us to deduce that the adsorbed hydrogen on the metal electrocatalyst acts as a poison and prevents the formation of the active intermediate lowering the reaction kinetics. Further the high activity on metal electrocatalysts confirm that the improved performance of $V^{2+}/V^{3+}$ is primarily due to metal itself, and not due to the change in carbon functionalization or increase in surface area. These findings can be utilized to develop more efficient electrocatalysts like metal alloys with ideal $\Delta G_H$ and $\Delta G_I$, thereby eliminating the $V^{2+}/V^{3+}$ kinetic overvoltage and allowing us to take a step forward in creating low cost VRFBs for a sustainable future. Additionally, the methodology used in this study to evaluate kinetic parameters and compare various metal electrocatalysts accurately can be extended to other redox couples employed in flow batteries like $Cr^{2+}/Cr^{3+}$, $Fe^{2+}/Fe^{3+}$, $Cu^{2+}/Cu^{+}$, etc.

REFERENCES:

(1) *DOE-EIA International Energy Outlook 2019.* (2019).

(2) NASA. Global Temperature | Vital Signs – Climate Change: Vital Signs of the Planet. (2021). Available at: https://climate.nasa.gov/vital-signs/global-temperature/. (Accessed: 28th November 2020)

(3) Bloomberg NEF. New Energy Outlook 2020. (2020).

(4) Dunn, B., Kamath, H. & Tarascon, J.-M. Electrical Energy Storage for the Grid : A Battery of Choices. *Sci. Mag.* **334**, 928–935 (2011).

(5) Sun, C.-N. *et al.* Probing Electrode Losses in All-Vanadium Redox Flow Batteries with Impedance Spectroscopy. *ECS Electrochem. Lett.* **2**, 43–45 (2013).

(6) **Agarwal, H.**, Florian, J., R. Goldsmith, B. & Singh, N. $V^{2+}/V^{3+}$ Redox Kinetics on Glassy Carbon in Acidic Electrolytes for Vanadium Redox Flow Batteries. *ACS Energy Lett.* **4**, 2368–2377 (2019).

(7) Li, B. *et al.* Bismuth Nanoparticle Decorating Graphite Felt as a High-Performance Electrode for an All-Vanadium Redox Flow Battery. *Nano Lett.* **13**, 1330–1335 (2013).

(8) Wei, L., Zhao, T. S., Zeng, L., Zhou, X. L. & Zeng, Y. K. Copper Nanoparticle-Deposited Graphite Felt Electrodes for All Vanadium Redox Flow Batteries. *Appl. Energy* **180**, 386–391 (2016).

(9) Mehboob, S. *et al.* Excellent Electrocatalytic Effects of Tin Through In Situ Electrodeposition on the Performance of All-Vanadium Redox Flow Batteries. *J. Mater. Chem. A* **5**, 17388–17400 (2017).

(10) Shen, J., Liu, S., He, Z. & Shi, L. Influence of Antimony Ions in Negative Electrolyte on the Electrochemical Performance of Vanadium Redox Flow Batteries. *Electrochim. Acta* **151**, 297–305 (2015).

(11) **Agarwal, H.**, Florian, J., Goldsmith, B. R. & Singh, N. The Effect of Anion Bridging on Heterogeneous Charge Transfer for $V^{2+}/V^{3+}$. *Cell Reports Phys. Sci.* **2**, 100307 (2021).

(12) Greeley, J., Jaramillo, T. F., Bonde, J., Chorkendorff, I. & Nørskov, J. K. Computational High-Throughput Screening of Electrocatalytic Materials for Hydrogen Evolution. *Nat. Mater.* **5**, 909–913 (2006).

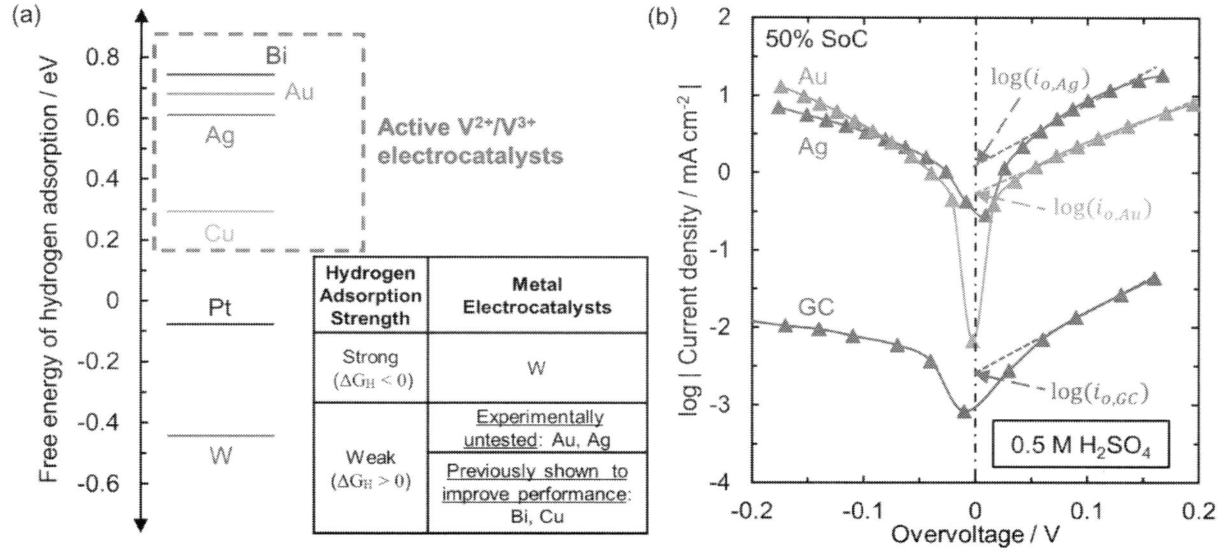


**Figure 1.** (a) Gibbs free energy of hydrogen adsorption ($\Delta G_H$) on metal surfaces at 1 bar $H_2$, values taken from ref. 12. The inset table shows metal electrocatalysts with varying $\Delta G_H$ and vanadium intermediate energy on which we tested our hypothesis that weakly adsorbed hydrogen facilitates intermediate formation by not competing for active sites in $V^{2+}/V^{3+}$ reaction, and (b) Log of absolute value of current densities vs overvoltage at 50% State of Charge $\left(\mathrm{SoC} = \frac{[V^{2+}]}{[V^{2+}]+[V^{3+}]}\right)$ in $H_2SO_4$ for Au and Ag metal electrocatalysts in comparison to glassy carbon (GC). The exchange current densities ($i_o$), evaluated by extrapolating $y$-intercept of log of oxidation current densities and using the Tafel equation for the reaction: $V^{2+} \rightarrow V^{3+} + e^-$, indicate that the activity for $V^{2+}/V^{3+}$ reaction on metal electrocatalysts Au and Ag (metals with positive $\Delta G_H$, shown in a) are ~two orders of magnitude higher than our benchmarking studies on GC.

# 632401 Non-Solvent Induced Phase Separation for Designer Porous Carbon Electrodes

**Abstract:**

Porous carbon electrodes play an important role in many leading electrochemical systems by supporting a number of critical functions which include providing active catalyst sites to facilitate redox reactions, distributing liquid reactants and products, and conducting electrons and heat.[1] This is particularly relevant for redox flow batteries (RFBs), wherein solubilized active species are forced through porous electrodes within a flow cell, oxidizing and reducing on the fiber surfaces during charge and discharge. Due to its decoupling of energy and power, scalability, and long service life, RFBs hold promise for long duration energy storage for the electric grid.[2] However, current RFBs are prohibitively expensive, necessitating further cost reduction for widespread deployment. Lowering reactor cost by improving power output is a promising approach to bridging the economic gap. While commercial electrodes are functional, they possess suboptimal surface chemistry and microstructure for existent and emerging RFB chemistries.[3] Deterministic fabrication of engineered porous materials would enable exploration of a wider electrode design space, further elucidating electrode-level performance descriptors.

In this presentation, we will describe a bottom-up synthesis method leveraging non-solvent induced phase separation (NIPS) for advancing porous electrodes with property sets suitable for RFBs.[4] NIPS enables fabrication of high surface area, interconnected pore networks unattainable in conventional fibrous electrodes; further, the pore size, gradient, and structure are tunable via facile synthesis design parameters. We demonstrate the feasibility of NIPS to generate a family of electrodes that share similar but uniquely distinct microstructural characteristics. Combining spectroscopic, microscopic, and physicochemical characterization to flow cell performance in two common aqueous RFB redox couples (i.e., iron chloride and all-vanadium), we will show the viability of this synthetic platform for illuminating structure-function relations in porous materials for RFBs. Further, we will highlight opportunities for the development of high-performance and chemistry-specific NIPS-derived electrode material sets. While the efforts of these studies are primarily focused on RFBs, we envision that the methods and findings would be beneficial for convection-driven electrochemical devices that require highly engineered electrodes.

**References:**

[1] Kim et al., *J. Mater. Chem. A.*, **3**, (2015).

[2] Dunn et al., *Science*, **334,** (2011)

[3] Forner-Cuenca et al., *Current Opinion in Electrochemistry*, **18**, (2019).

[4] Wan & Jacquemond et al., *Adv. Mat.*, (2021).

# 623543 The Role of Water in Vapor-Fed Proton-Exchange-Membrane Electrolysis

**Abstract:**

Electrolytic devices provide a carbon-free option for producing hydrogen, with many advantages such as modular, scalable design and flexibility of operating parameters (temperature, pressure, etc.). Typically, electrolyzers are run under conditions that require a significant amount of clean, liquid water and are run at several $A/cm^2$. Using liquid water mitigates transport issues at these high current densities, but introduces multiphase flow when liquid water and oxygen gas are present simultaneously. These challenges can be surmounted using a vapor-fed electrolyzer, where water vapor is split to oxygen and hydrogen directly. While removing multiphase flow, vapor-fed electrolysis introduces a myriad of other considerations when investigating the water, ion, and gas transport within the system.

In this talk, we explore water vapor electrolysis, both with experiments and with a mathematical model, to elucidate the local transport in the cell. Applied voltage breakdowns are used to analyze the losses are occurring and how they differ from the transport in liquid-phase systems. The major limitations within the vapor-fed electrolyzer are caused by a decreased water content within the membrane phase, indicated by increased Ohmic and mass transport losses seen in applied voltage breakdowns. Detailed explanation of other transport, including the water within the gas phase and in the membrane, provides a further understanding of transport limitations for these vapor-fed cells. The performance, though limited compared to liquid systems, has the potential to satisfy niche applications for future study.

J.C.F. would like to acknowledge support from National Science Foundation Graduate Research Fellowship under Grant No. DGE 1106400. This work was partially funded by the Energy & Biosciences Institute through the EBI-Shell program.

frontiers in Energy Research

ORIGINAL RESEARCH
published: 15 October 2021
doi: 10.3389/fenrg.2021.725010

# Energetics of $Li^+$ Coordination with Asymmetric Anions in Ionic Liquids by Density Functional Theory

*Drace Penley[1], Stephen P. Vicchio[2], Rachel B. Getman[2] and Burcu Gurkan[1]**

[1]Chemical and Biomolecular Engineering, Case Western Reserve University, Cleveland, OH, United States, [2]Chemical and Biomolecular Engineering, Clemson University, Clemson, SC, United States

The energetics, coordination, and Raman vibrations of Li solvates in ionic liquid (IL) electrolytes are studied with density functional theory (DFT). $Li^+$ coordination with asymmetric anions of cyano(trifluoromethanesulfonyl)imide ([CTFSI]) and (fluorosulfonyl)(trifluoro-methanesulfonyl)imide ([FTFSI]) is examined in contrast to their symmetric analogs of bis(trifluoromethanesulfonyl)imide ([TFSI]), bis(fluorosulfonyl)imide ([FSI]), and dicyanamide ([DCA]). The dissociation energies that can be used to describe the solvation strength of $Li^+$ are calculated on the basis of the energetics of the individual components and the Li solvate. The calculated dissociation energies are found to be similar for $Li^+$-[FTFSI], $Li^+$-[TFSI], and $Li^+$-[FSI] where only $Li^+$-O coordination exists. Increase in asymmetry and anion size by fluorination on one side of the [TFSI] anion does not result in significant differences in the dissociation energies. On the other hand, with [CTFSI], both $Li^+$-O and $Li^+$-N coordination are present, and the Li solvate has smaller dissociation energy than the solvation by [DCA] alone, [TFSI] alone, or a 1:1 mixture of [DCA]/[TFSI] anions. This finding suggests that the $Li^+$ solvation can be weakened by asymmetric anions that promote competing coordination environments through enthalpic effects. Among the possible Li solvates of $(Li[CTFSI]_n)^{-(n-1)}$, where $n = 1, 2, 3$, or 4, $(Li[CTFSI]_2)^{-1}$ is found to be the most stable with both monodentate and bidentate bonding possibilities. Based on this study, we hypothesize that the partial solvation and weakened solvation energetics by asymmetric anions may increase structural heterogeneity and fluctuations in Li solvates in IL electrolytes. These effects may further promote the $Li^+$ hopping transport mechanism in concentrated and multicomponent IL electrolytes that is relevant to Li-ion batteries.

**Keywords: ionic liquid electrolytes, lithium solvation, concentrated electrolytes, lithium-ion battery, *ab initio* thermodynamics**



***Edited by:***
*Ah-Hyung Alissa Park, Columbia University, United States*

***Reviewed by:***
*Hailei Zhao, University of Science and Technology Beijing, China*
*Shrihari Sankarasubramanian, University of Texas at San Antonio, United States*

****Correspondence:***
*Burcu Gurkan*
*beg23@case.edu*

***Specialty section:***
*This article was submitted to Electrochemical Energy Conversion and Storage, a section of the journal Frontiers in Energy Research*

***Received:*** *14 June 2021*
***Accepted:*** *10 September 2021*
***Published:*** *15 October 2021*

***Citation:***
*Penley D, Vicchio SP, Getman RB and Gurkan B (2021) Energetics of Li⁺ Coordination with Asymmetric Anions in Ionic Liquids by Density Functional Theory. Front. Energy Res. 9:725010. doi: 10.3389/fenrg.2021.725010*

## INTRODUCTION

The structure of $Li^+$ solvates in concentrated electrolytes has a significant impact on $Li^+$ transport and is influential in the rate capability of rechargeable lithium ion batteries (LIBs) (Borodin et al., 2018; Yamada et al., 2019; Krachkovskiy et al., 2020; Pham et al., 2021). ILs present an extreme case of a concentrated electrolyte where the electrolyte is made entirely of discrete ions and lacks neutral solvent molecules. Generally, ILs are known to have high ionic conductivity, large electrochemical windows, and negligible volatility (Bae et al., 2013; Navarra, 2013). These properties are desirable for lithium battery systems and thus are of great interest for safe, high-density energy storage devices (Galinski et al., 2006; Bae et al., 2013; Navarra, 2013; Eftekhari et al., 2016). However, $Li^+$ transport in

ILs is complex and hindered by the high viscosity, mainly due to Coulombic interactions and specific $Li^{+}$–anion interactions, which leads to the formation of clusters that decrease the mobility of $Li^{+}$ in the bulk liquid (Lesch et al., 2014; Pham et al., 2021). $Li^{+}$ transport in ILs has been studied mostly by Raman spectroscopy, nuclear magnetic resonance spectroscopy, transference measurements, density functional theory (DFT), and classical molecular dynamic (MD) simulations (Borodin et al., 2006; Umebayashi et al., 2007; Duluard et al., 2008; Lassègues et al., 2009; Umebayashi et al., 2010; Castiglione et al., 2011; Fujii et al., 2013; Haskins et al., 2014; Borodin et al., 2018). In this study, the DFT approach is employed to understand the impact of anion asymmetry and the presence of binary anions on the structure and energetics of the first solvation shell of $Li^{+}$ in IL electrolytes.

The bis(trifluoromethanesulfonyl)imide, [TFSI], anion and its respective ILs have been widely studied as potential electrolytes for lithium metal batteries and LIBs, and were shown to improve battery performance over the state-of-the-art organic carbonate electrolytes with Li[$PF_6$] (Sakaebe and Matsumoto, 2003; Garcia et al., 2004; Armand et al., 2009). Fundamental studies of IL electrolytes with [TFSI] anion showed that $Li^{+}$ coordinates with anionic oxygen atoms where on average slightly less than four oxygen atoms are involved (Borodin et al., 2006), leading to two anions in the first solvation shell where *cis* conformation is favored with increased Li salt doping (Umebayashi et al., 2010). These studies were focused on Li salt mole fraction of $x_{Li^{+}}<0.2$. In the similar Li salt concentration range, Fujii et al. showed that in the case of [FSI], the solvation number increases to three anions in the first shell where both monodentate and bidentate bonds coexist (Fujii et al., 2013; Lesch et al., 2016). Later, Lesch et al. (2016) showed that when both [TFSI] and [FSI] anions are available, $Li^{+}$ prefers to coordinate with [TFSI] and forms $Li^{+}$-[TFSI]-$Li^{+}$ dimers which move on a similar timescale as $Li^{+}$-[TFSI] aggregates. Haskins et al. (2014) studied the $Li^{+}$ solvation and transport in pyrrolidinium ILs with [TFSI] and [FSI] anions, and imidazolium IL with tetrafluoroborate [$BF_4$] anion, with each system having 0.5 molal Li-salt concentration. When examining the solvation structures, they observed that the binding distances increased in the order of $Li^{+}$-[FSI] > $Li^{+}$-[TFSI] > $Li^{+}$-[$BF_4$], where the system with [$BF_4$] had the strongest binding energy and the system with [FSI] had the weakest binding energy. Different than previous studies, Haskins et al. reported that three [TFSI] anions coordinate with $Li^{+}$ through two monodentate and two bidentate bonds, while the coordination with [$BF_4$] and [FSI] involved four anions with each having monodentate binding. The differences from the previous reports were attributed to the higher Li salt concentrations in their study ($x_{Li^{+}}\sim0.33$) and difficulty in characterizing the coordination environment with experimental spectroscopy in the previous reports. At high Li salt concentrations, Li et al. (2012) also showed increased [TFSI] coordination with fewer bidentate and more monodentate bonds. Haskins et al. also studied the residence time of neighboring ions to $Li^{+}$ among other transport parameters, including $Li^{+}$ diffusivity and conductivity. The correlated ion motion that corresponds to the vehicular motion was decreased with increasing Li salt concentration, suggesting that $Li^{+}$ moves through the liquid by anion exchange. They concluded that the majority (60%) of the transport occurred *via* vehicular motion, while $Li^{+}$ hopping was found to gain more significance with large-sized anions and increased $Li^{+}$ concentration. Therefore, the promotion of the hopping mechanism *via* anion exchange is the key for enhanced $Li^{+}$ mobility in concentrated electrolytes such as ILs considering the limitations of the vehicular mobility in the presence of $Li^{+}$ aggregates at $x_{Li^{+}}>0.05$ (Brinkkötter et al., 2018; McEldrew et al., 2021).

Decreasing the solvation strength and the lifetime of coordinated species in LIB electrolytes have been suggested to play an important role in improving $Li^{+}$ mobility. Borodin et al. (2017) demonstrated the promotion of $Li^{+}$ transport in concentrated electrolytes where nano-heterogeneity decouples cations from anion cages. In other examples where heterogeneity was achieved by simple mixtures, $Li^{+}$ solvation was shown to weaken in binary organic carbonate mixtures (Matsuda et al., 2002), and the solvation shell was found to be more rigid in cyclic carbonates than in their linear counterparts (Fulfer and Kuroda, 2016). Shim (2018) studied the solvation structure and dynamics of a mixture of ethylmethyl carbonate (EMC), dimethyl carbonate (DMC), and ethylene carbonate (EC) with the Li[$PF_6$] salt by MD simulations. They reported that as the EC concentration increased, the solvent fluctuations in Li solvates decreased due to the preferred coordination of $Li^{+}$ with EC and increased local density by the ring structure of EC, which was consistent with the reports by Yang et al. (2010). The decreased solvent fluctuations resulted in increased viscosity and reduced $Li^{+}$ diffusion. Similar to organic mixture electrolytes, mixtures of ILs have been studied to increase conductivity by Li solvate structural fluctuations and reduced solvation strength (Yang et al., 2010; Kerner et al., 2015; Huang et al., 2019a; Huang et al., 2019b; Chen and Forsyth, 2019).

IL mixtures with [TFSI]/[FSI] anions have been studied the most to date due to the superior electrochemical properties of ILs paired with these anions. The [DCA] anion has also garnered attention due to its lower viscosity and higher conductivity than [TFSI]- and [FSI]-based ILs. In a study by Yoon et al. (2013), the [DCA] anion paired with a pyrrolidinium cation in a Li|$LiFePO_4$ cell showed stable cycling at elevated temperatures. Huang et al. (2019b) showed by a combined Raman spectroscopy and MD simulations study that at $x_{Li^{+}}<0.2$, $Li^{+}$ coordinates on average with four [DCA] anions *via* monodentate bonds in a binary Li salt/IL mixture with [DCA] anion as well as a Li salt/IL/IL ternary mixture with [TFSI]/[DCA] anions. Using attenuated total reflectance far-ultraviolet and deep-ultraviolet spectroscopy, Imai et al. (2020) reported similar results for the first solvation shell of $Li^{+}$ in the Li salt/IL binary electrolyte with [DCA] anions. Huang et al. (2019b) also studied the ion pair lifetimes by MD and $Li^{+}$ transference by an experimental polarization method employed in Li-Li symmetrical cells. It was found that $Li^{+}$ coordination was not a strong function of the Li salt in the ternary mixture with [TFSI]/[DCA] anions, in contrast to Li-salt/IL binary electrolytes with [TFSI]. The ion pair lifetime of $Li^{+}$-[TFSI] decreased with increasing [TFSI] concentration in the ternary electrolyte. This result was attributed to $Li^{+}$ preferring to coordinate with [DCA] and the addition of the [TFSI] anion breaking down the rigid first solvation shell due to the weaker coordination bond.

**TABLE 1 |** Anions of interest with full name as well as an abbreviation used within the study.

| Anion | Full name | Structure |
|---|---|---|
| [DCA] | Dicyanamide | 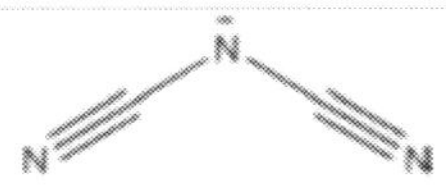 |
| [CFSI] | Cyano(fluorosulfonyl)imide | 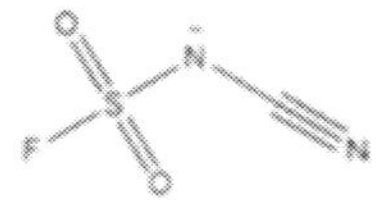 |
| [CTFSI] | Cyano(trifluoromethanesulfonyl) imide | 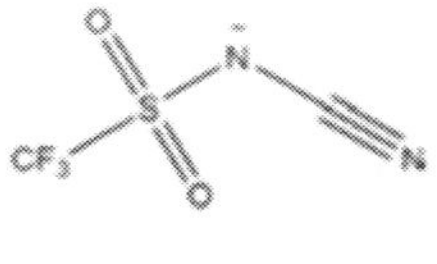 |
| [FSI] | Bis(fluorosulfonyl)imide | 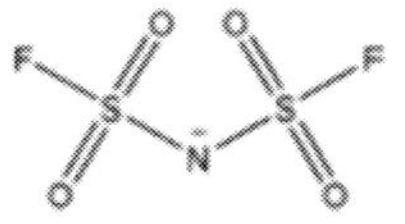 |
| [TFSI] | Bis(trifluoromethanesulfonyl) imide | 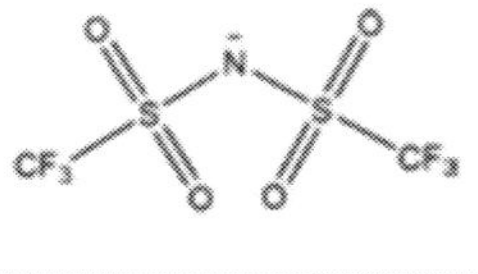 |
| [FTFSI] | (fluorosulfonyl) (trifluoromethanesulfonyl) imide | 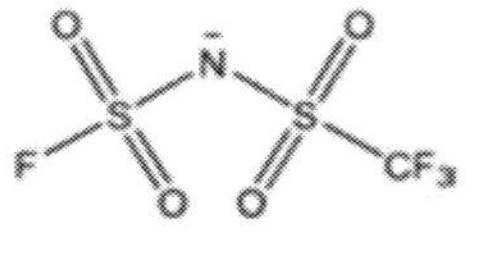 |
| [PFTFSI] | (pentafluoroethanesulfonyl) (trifluoromethanesulfonyl)imide | 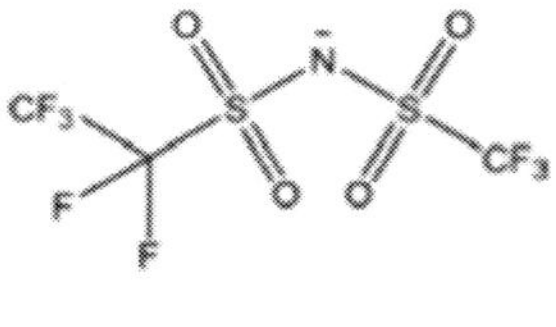 |
| [HFTFSI] | (heptafluoropropanesulfonyl) (trifluoromethanesulfonyl)imide | 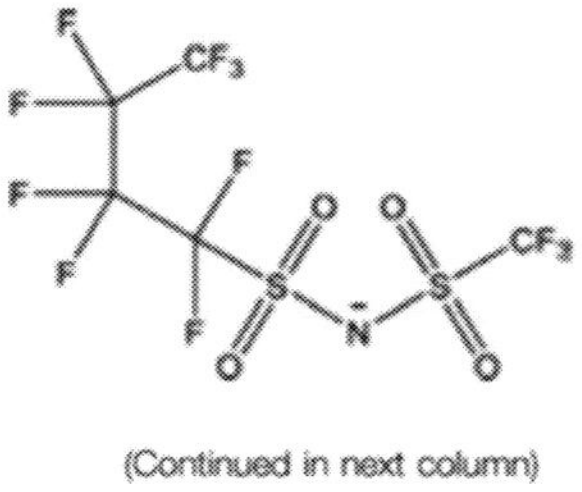 |
| [NFFSI] | (nonafluorobutanesulfonyl) (fluorosulfonyl)imide | 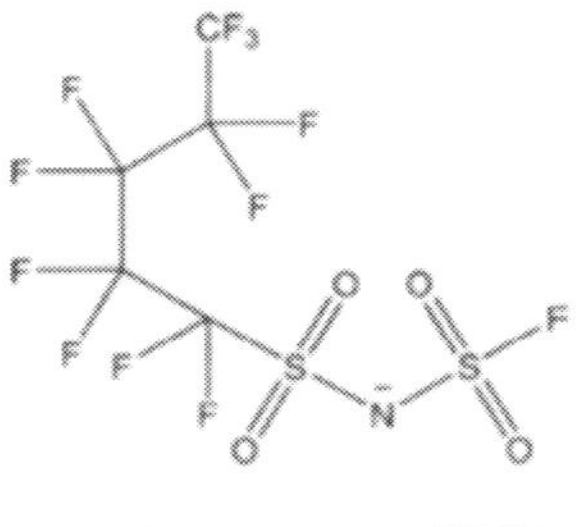 |
| [NFTFSI] | (nonafluorobutanesulfonyl) (trifluoromethanesulfonyl) imide | 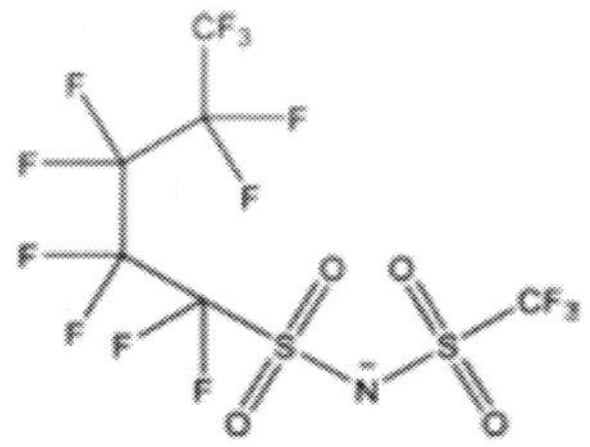 |

(Continued in next column)

**TABLE 1 |** (*Continued*) Anions of interest with full name as well as an abbreviation used within the study.

Very recently, Nurnberg et al. (2020) studied the $Li^+$ solvation and transport in IL electrolytes ($x_{Li^+}$ = 0.1–0.7) with the [CTFSI] anion by Raman spectroscopy and pulsed gradient NMR measurements. This anion is an asymmetric analog to both [TFSI] and [DCA] anions. They found that the [CTFSI]-based electrolyte had lower conductivity than the [TFSI] system. However, a pronounced transition from vehicular to structural diffusion (hopping) was noted for the asymmetric anion. Studies on such asymmetric anions are scarce, and mixture anions in IL electrolytes are relatively recently being explored. Understanding the $Li^+$ coordination environments and the solvation energies in the case of asymmetric anions are important for tuning concentrated electrolytes for future energy storage devices. In this study, we calculated the optimized geometries, Raman vibrations, and the Li solvate energies by DFT to understand the effect of anion functionality in IL mixtures with Li salts. Several asymmetric anions based on [FSI], [TFSI], and [DCA] parent anions were studied and compared in terms of the anion size, extent of fluorination, and coordination chemistry. The structures of the studied anions as well as their abbreviation are listed in **Table 1**.

The bulk properties and electrochemical applications of the three asymmetric anions such as [NFTFSI], [CTFSI], and [FTFSI] are relatively unexplored when compared to their parent anions. The [CTFSI] anion in an IL electrolyte has shown promise in LIBs, with improved thermal and anodic stability, and exhibited higher room temperature conductivity than [DCA] and [TFSI] parent anions (Hoffknecht et al., 2017). However, the structure of the Li solvate was not studied, and it is not clear what the role of the liquid structure is on the improved performance. An investigation on the [FTFSI]-based IL electrolyte showed that

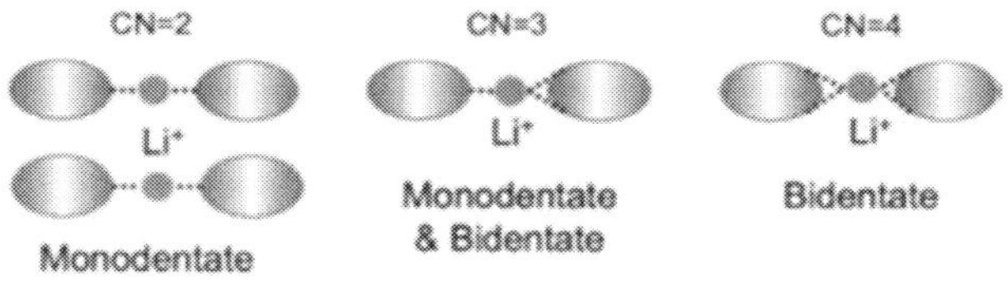


**FIGURE 1** | Representative coordination environments for $Li^+$ with two asymmetric solvating anions forming bidentate and/or monodentate bonds. $Li^+$ is represented by the purple circle and the asymmetric anion is represented by the ovals, with red and blue color codes corresponding to the different coordination sites.

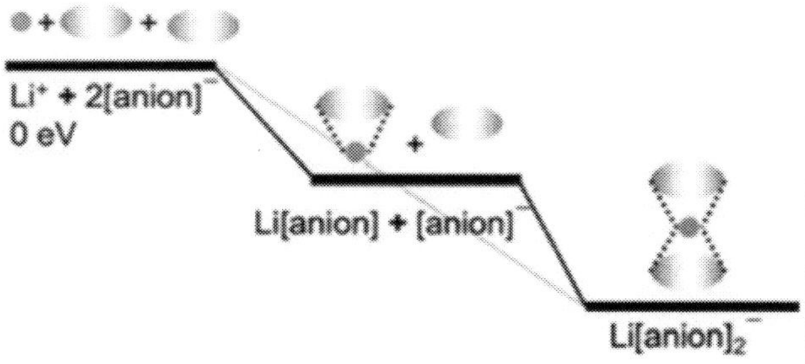


**FIGURE 2** | Comparison of differential and integral analyses for Li solvates with a symmetrical anion. Bidentate binding is illustrated for one and two coordinating anions. Energy levels are indicated by the black horizontal bars where the starting individual components have 0 eV as a reference and the Li solvates have <0 eV. The differential energies are represented by the solid black lines and the integral energy is represented by the diagonal green line that connects the 0 eV state with the $Li[anion]_2^{-1}$ solvate at the lowest energy state.

the vehicular motion dominates due to the formation of $Li^+$ aggregates until 0.4 mole fraction of the Li salt after which the transport of $Li^+$ *via* hopping increases (Brinkkötter et al., 2018). [NFTFSI] is unique in that rather than combining the functionality of two parent IL anions, the fluoromethane group on [TFSI] is replaced with a fluorobutane. This provides the ability to quantify how the increased fluorination impacts the structure and transport properties. There has been little investigation of the [NFTFSI] and [NFFSI] anions in the past. Palumbo et al. (2017) studied the conformer structures of [NFTFSI] in a pyrrolidinium IL. Han et al. (2011) utilized 1M Li[NFFSI] as the conducting salt in organic carbonates (EC/EMC 3:7 v/v) to improve high temperature performance of LIBs. Li [NFFSI] showed similar conductivity as $Li[ClO_4]$ and slightly lower than $Li[PF_6]$; however, Li[NFFSI] did not corrode the aluminum current collector of the graphite/$LiCoO_2$ cell with 63% capacity retention over 100 cycles at 60°C, while the cell with $Li[PF_6]$ failed rapidly. Therefore, how the length of fluorination impacts the stability of LIBs is important to understand for extreme temperature conditions. In order to do so, we need a better understanding of how these fluorinated anions solvate and interact with $Li^+$ in the bulk electrolyte. The [CFSI], [PFTFSI], and [HFTFSI] anions were studied here to explore how the degree of fluorination of the asymmetric anions impacts solvation strength. Previously, it was found that the binding energies of Li solvates with organic ligands decreased with increased fluorination (Bauschlicher, 2018). It was also shown that with reduced dipole moment and an increased ligand size, the ligand–ligand repulsion was increased. This was discussed to possibly lead to different structures than the non-fluorinated analogs. In this study, we show how the strength of solvation is impacted by the asymmetry of the coordinating anions and how the anion mixtures compare to the analogous asymmetric anion that has binary coordination with $Li^+$. We report the strength of the first solvation shell of $Li^+$ in ILs without the impacts of the IL cation through DFT calculations and suggest which anions may help promote $Li^+$ transport based on this first principles study.

# METHODS

## Computational Details

Dissociation energies ($E_D$) were used to assess the thermodynamic favorabilities of the different $Li^+$-anion systems. $E_D$ represents the amount of energy it takes to break $Li^+$-anion clusters into their constituents (i.e., the difference in energy going from the $Li^+$-anion cluster to the individual components of $Li^+$ and the anion at infinite separation). They were calculated using quantum mechanics in the absence of solvation, similar to a prior study on $Li^+$ transport in cross-linked polymer electrolytes with ILs (Elmore et al., 2018), according to the equation:

$$E_D = \left[E_{Li^+} + \sum_i y(E_{anion})_i\right] - E_{Li\text{-}solvate}, \quad (1)$$

where $E_{Li}{}^+$ is the electronic energy of the lithium ion, $E_{anion}$ is the electronic energy of the isolated anion $i$, $y$ is the number of anion $i$ in the solvate structure, and $E_{Li\text{-}solvate}$ is the electronic energy of the Li solvate. In this study, $1 \le y \le 4$. The isolated $Li^+$ is modeled with a +1 charge, and each anion is modeled with a net charge of −1. The net charge of the Li solvate can hence be 0, −1, −2, and −3, depending on the number of coordinating anions. This analysis is limited to a maximum number of coordinating anions of four due to the repulsive forces between anions when the first solvation shell includes five anions or greater, resulting in concentrated environments of $0.20 \le x_{Li}{}^+ \le 0.50$. According to **Eq. 1**, positive values of $E_D$ are favorable, with increasing values of $E_D$ indicating stronger $Li^+$ ion bonds. Thus, a more positive $E_D$ value is interpreted as a more stable Li solvate, and a more negative $E_D$ value is interpreted as less stable, which would allow for faster breakdown of the first solvation shell and thus quicker exchange of anions within this shell.

Quantum mechanics calculations were carried out in the Gaussian 16 program (Frisch et al., 2016) using the three-parameter Becke model with the Lee–Yang–Par modification (B3LYP) (Lee et al., 1988) functional for exchange and correlation and the 6-311G(d,p) (Ditchfield et al., 1971) basis set. Structures of the isolated anions and $Li^+$ solvate systems and their associated electronic energies were obtained using geometry optimizations with default convergence criteria; as the $Li^+$ system does not require geometry optimization, its electronic energy was calculated as a "single point." Frequency calculations were carried out at the same level of theory in order to obtain thermodynamic and Raman spectroscopic data. Raman frequencies were scaled by a factor of 0.964 (Palafox, 2018). An example Gaussian input file for a geometry optimization and frequency calculation is provided in the supplementary.

## Coordination Number and Li Solvate Structures

It is important to note the distinction between the number of anions coordinating with $Li^+$ and the coordination number (CN) associated with this first solvation shell. An anion can have more than one bonding site for $Li^+$ in which case CN will be a greater number than the number of coordinating anions. For example, the *cis* and *trans* conformation of [TFSI] can yield bidentate and monodentate bonds which can result in CN of 1–4 as [TFSI] has four oxygen atoms that can coordinate with $Li^+$. **Figure 1** demonstrates the possible monodentate and bidentate binding with two coordinating anions. The represented anions are asymmetric as illustrated by the color code (blue and red). Thus, with two coordinating asymmetric anions, a CN of two, three, or four is possible.

## Differential vs. Integral Analysis

Two methods were used to determine the energy difference associated with adding a new anion to the first solvation shell of $Li^+$. These methods can be visualized in **Figure 2** for a symmetric anion such as [TFSI] coordination with $Li^+$. The integral binding energy, which is equal to $-1 \times E_D$, is shown *via* the green diagonal line. It refers to the energy to add *n* anions to $Li^+$, for example, $Li^+ + 2[\text{anion}]^- \rightarrow Li[\text{anion}]_2^-$. In contrast, the differential binding energy is the energy to add one anion at a time, for example, $Li[\text{anion}] + [\text{anion}]^- \rightarrow Li[\text{anion}]_2^-$. The differential binding energy is useful for assessing the maximum number of anions that will coordinate to $Li^+$, whereas the integral binding energy is useful for assessing relative stabilities. A detailed example of the asymmetric [CTFSI] and [FTFSI] differential vs. integral analysis can be seen in **Supplementary Tables S1, S2**.

## *Ab Initio* Thermodynamic Analysis

The thermodynamic stabilities of structurally and compositionally different Li solvates were calculated using an *ab initio* thermodynamic analysis (Soon et al., 2007; Getman et al., 2008; Grundner et al., 2015; Paolucci et al., 2016). This analysis was performed specifically for [TFSI]- and [DCA]-containing solvates. The *ab initio* thermodynamic analysis calculates the free energy of formation for different Li solvates at different mole fractions of [DCA] and [TFSI] in solution according to the following equation:

$$Li^+ + a[DCA] + b[TFSI] \overset{\Delta G}{\leftrightarrow} Li[DCA]_a[TFSI]_b, \quad (2)$$

where *a* is the number of [DCA] anions and *b* is the number of [TFSI] anions that coordinate to a single $Li^+$ cation. We used a library for the analysis that includes all [DCA]-only ($1 \le a \le 4$; b = 0), [TFSI]-only (a = 0; $1 \le b \le 3$), and [DCA]/[TFSI] ($1 \le a + b \le 3$) structures. According to the *ab initio* thermodynamic analysis, the free energy of formation for compositionally different structures ($Li[DCA]_a[TFSI]_b$) is calculated by transforming the free energy of a fixed number of species, $\Delta G(T, P, N_{Li^+}, N^*_{[TFSI]}, N^*_{[DCA]})$, where $*$ denotes that the anion species is coordinated with the $Li^+$ cation, to a fixed chemical potential using Legendre's transformations (see Supporting Information for more details). In this study, we transform $N^*_{[TFSI]}$ and $N^*_{[DCA]}$ to $\mu^*_{[TFSI]}$ and $\mu^*_{[DCA]}$; the resulting free energy of formation is denoted as $\Delta G^{(2)}$ (where the 2 superscript denotes two variables were transformed) and is given by the following equation:

$$\Delta G^{(2)}(T, P, N_{Li^+}, \mu^*_{TFSI}, \mu^*_{DCA}) = (\Delta E) - (\mu^*_{TFSI})(N^*_{TFSI}) - (\mu^*_{DCA}) \times (N^*_{DCA}), \quad (3)$$

where $\Delta E$ is the difference in electronic energy between a $Li[DCA]_a[TFSI]_b$ structure and the isolated $Li^+$ cation. Chemical potentials of [DCA] and [TFSI] are calculated (Starzak, 2010; Moučka et al., 2015) as follows:

$$\mu^*_i(T, x_i) = E_i + RTln(\gamma_i x_i), \quad (4)$$

where $E_i$ is the electronic energy of the anion (*i* = DCA or TFSI), $x_i$ is the mole fraction in solution, *R* is the gas constant, and $\gamma_i$ is the activity coefficient. We assume a temperature of 298 K for all calculations and that all activity coefficients are equal to unity (Noda et al., 2001; Sun et al., 2018). The influence of this choice is detailed in the Supporting Information.

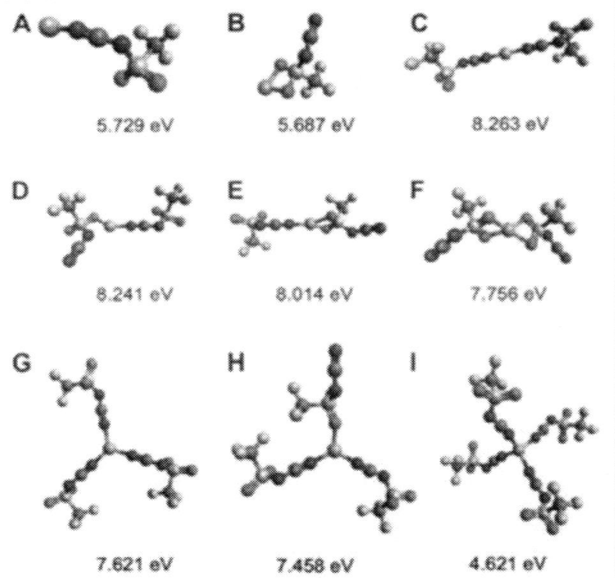


**FIGURE 3** | Geometry optimized structures and calculated integral energies, $E_D$, of Li solvates with [CTFSI] for the number of coordinating anions of 1 **(A,B)**, 2 **(C–F)**, 3 **(G,H)**, and 4 **(I)**. The atom color code is as follows: red (oxygen), blue (nitrogen), yellow (sulfur), gray (carbon), purple (lithium), and light blue (fluorine).

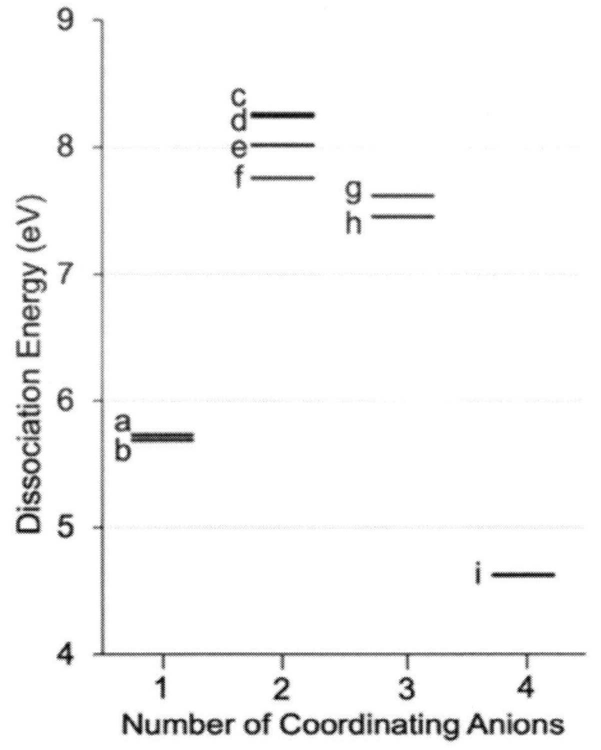


**FIGURE 4** | Integral dissociation energies, $E_D$, for Li solvates with [CTFSI] according to the numbers of coordinating anions. Labels **a–i** correspond to structures shown in **Figure 3**.

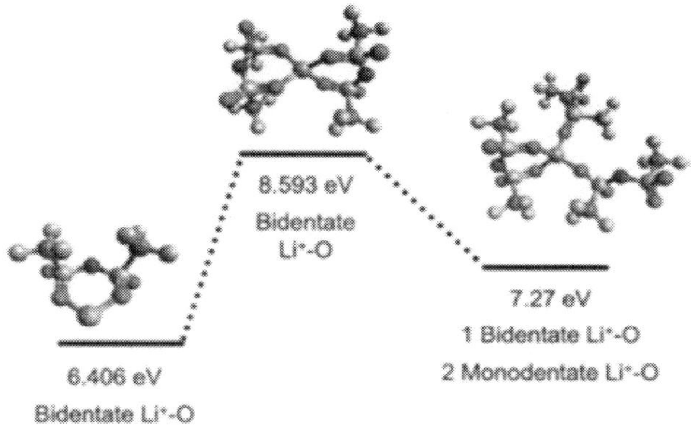


**FIGURE 5** | Integral disassociation energies, $E_D$, and representative structures for $Li^+$-[TFSI] solvates with the respective coordination type. The color code for atoms is as follows: red (oxygen), blue (nitrogen), yellow (sulfur), gray (carbon), purple (lithium), and light blue (fluorine).

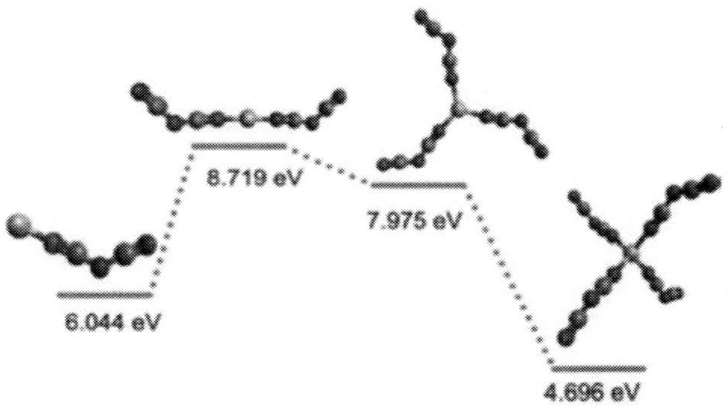


**FIGURE 6** | Integral dissociation energies, $E_D$, and representative structures for $Li^+$-[DCA] systems with the respective coordination type (all monodentate $Li^+$-N). The color code for the atoms is as follows: blue (nitrogen), gray (carbon), and purple (lithium).

## RESULTS AND DISCUSSION

### Thermodynamically Favored Structure

The $Li^+$ solvation in the electrolyte is dynamic, and different populations of multiple solvate structures maybe present. In order to determine the most likely configurations, dissociation energies were calculated for all of the possible monodentate and bidentate coordination environments while incrementally increasing the number of anions around $Li^+$. The list of calculated $E_D$'s for all of the Li solvates studied is in **Supplementary Table S3**. For the asymmetric [CTFSI], it is determined that $Li^+$ can coordinate with up to four anions where CN varies from one to four, as shown in **Figure 3**. Both the oxygen on the sulfonyl ($Li^+$-O) and the nitrogen on the cyano group ($Li^+$-N) can coordinate with $Li^+$. Therefore, when there is one [CFTSI] coordinating with $Li^+$, two configurations are possible, as seen in **Figures 3A,B**. With two coordinating anions, there are four possible configurations, as seen in **Figures 3C–F**. With three [CTFSI] anions, there are only two possible structures: all cyano coordination with monodentate bonds or two cyano and one sulfonyl oxygen coordination each with monodentate bonds, as shown in **Figures 3G,H**. Finally, moving to four coordinating anions, the only possible structure is seen in **Figure 3I** *via* monodentate $Li^+$-N bonds. The calculated $E_D$'s for these Li solvate structures at differing numbers of coordinating anions are shown in **Figure 4**. It is interpreted that the structure with the highest $E_D$ is the most stable since it requires more energy to break the coordination. Therefore, Li solvates with the highest $E_D$ (strongest) among the multiple possible configurations for a specific coordination are said to be thermodynamically favored. According to the energies in **Figure 4**, for both Li[CTFSI] and $(Li[CTFSI]_2)^{-1}$, $Li^+$ prefers to coordinate with [CTFSI] *via* the cyano group ($Li^+$-N) over the sulfonyl oxygen atoms ($Li^+$-O). The most optimal Li solvate is $(Li[CTFSI]_2)^{-1}$ where $Li^+$ coordinates with two cyano groups each through monodentate conformation as in **Figure 3C** and/or one monodentate with cyano and another monodentate with sulfonyl, as in **Figure 3D**. Based on the three and four anion systems relying solely on $Li^+$-N coordination, all $Li^+$-N coordinated $(Li[CTFSI]_2)^{-1}$ are used for further analysis. This methodology for selecting the thermodynamically favored structures was implemented for all of the anions investigated in this study.

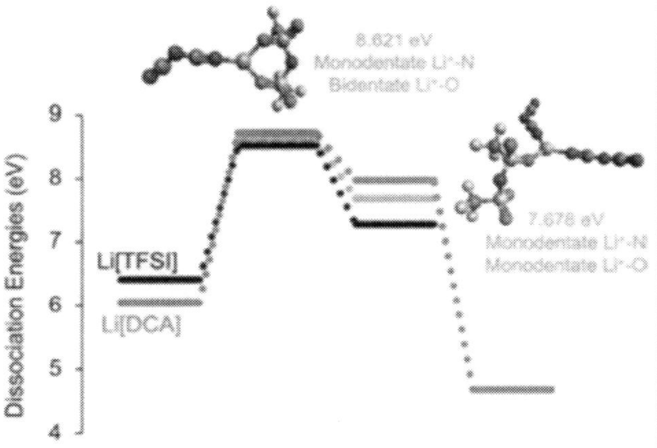


**FIGURE 7** | Integral dissociation energies, $E_D$, and representative structures for Li solvates with [DCA]/[TFSI] anions (orange bars) in comparison to Li-[TFSI] (black bars) and Li-[DCA] (blue bars). The color code for the atoms is as follows: red (oxygen), blue (nitrogen), yellow (sulfur), gray (carbon), purple (lithium), and light blue (fluorine).

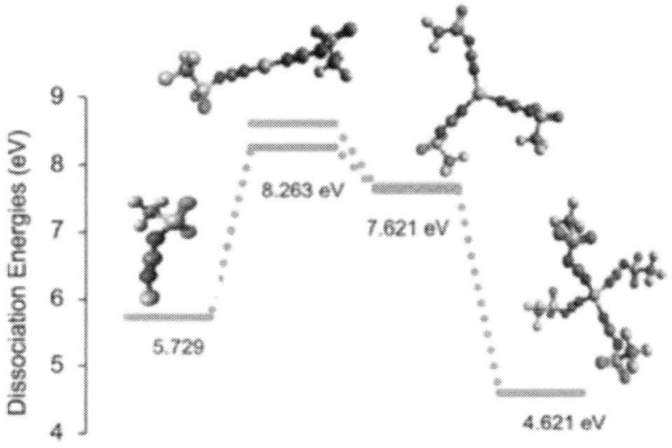


**FIGURE 8** | Integral dissociation energies, $E_D$, and representative structures for Li-[CTFSI] solvates (green bars) in comparison to the Li-[DCA]/[TFSI] mixture (orange bars). All of the coordination environments are based on monodentate Li-N bonds. The color code for the atoms is as follows: red (oxygen), blue (nitrogen), yellow (sulfur), gray (carbon), purple (lithium), and light blue (fluorine).

To better understand the change in solvation energy with the asymmetric anion in comparison to the symmetric anions, the $Li^+$-[CTFSI] system was compared with the $Li^+$-[TFSI] and $Li^+$-[DCA] which are more widely studied experimentally (Sakaebe and Matsumoto, 2003; Garcia et al., 2004; Borgel et al., 2009; Yoon et al., 2013; Shen et al., 2015).

Further, the Raman frequencies of [TFSI], [DCA], and [CTFSI] with and without $Li^+$ coordination were calculated (spectra shown in **Supplementary Figures S1–S3**). The prominent vibrations were of the S-N-S and S=O bonds of [TFSI], and the C-N-C bond of [DCA]. For [CTFSI], the vibrations of interest are S-N-$C_{CN}$, S=O, and C≡N. With $Li^+$ coordination, the symmetric S-N-S stretch of *cis*-[TFSI] at 735 $cm^{-1}$ experiences a blue shift by 10 $cm^{-1}$, which is consistent with previously reported experiments (Huang et al., 2019a; Huang et al., 2019b; Nurnberg et al., 2020). The S=O symmetric stretch appears at 1,132 and 1,148 $cm^{-1}$. With $Li^+$ coordination (structure in **Figure 5** with single [TFSI] coordinated to $Li^+$), the S=O peaks shift to 1,100 and 1,111 $cm^{-1}$. These vibrations have similar trends with [CTFSI]. Specifically, the S-N-$C_{CN}$ peak in [CTFSI] at 762 $cm^{-1}$ shifts to 779 $cm^{-1}$ with $Li^+$ coordination (structure in **Figure 3B**). The S=O vibration of [CTFSI] at 1,118 $cm^{-1}$ shifts to 1,099 $cm^{-1}$. Similarly, the C-N-C asymmetric and symmetric stretches of [DCA] at 2,183 and 2,199 $cm^{-1}$ show blue shifts by 15 and 45 $cm^{-1}$ (structure in **Figure 6** with single [DCA] coordinated to $Li^+$), respectively, which are also consistent with previous experimental findings (Huang et al., 2019b). [CTFSI] shows a similar peak location as [DCA] for the C≡N symmetric stretch at 2,195 $cm^{-1}$, with a blue shift of 42 $cm^{-1}$ with $Li^+$ coordination (structure in **Figure 3A**), which is similar to the previous experimental findings (Nurnberg et al., 2020). The overall consistency of the peak locations and shifts when moving from the pure anion to the Li solvate provide confidence that our converged structures closely match with the known experimental reports.

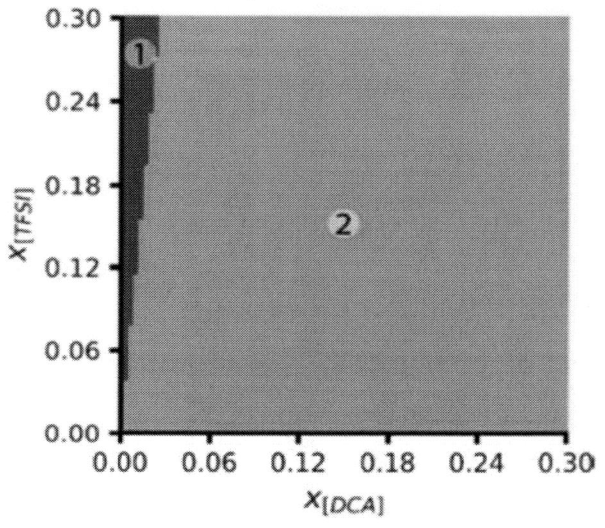


**FIGURE 9** | Phase diagram for $Li^+$ solvates calculated in the *ab initio* thermodynamic analysis as a function of the mole fractions of [DCA] and [TFSI]. Color key: the purple region (1) = $(Li[TFSI]_2)^{-1}$ and the orange region (2) = $(Li[DCA]_2)^{-1}$.

## $Li^+$ Solvation by [TFSI], [DCA], and [CTFSI]

**Figures 5, 6** show the dissociation energies and structures for Li solvates with the [TFSI] and [DCA] anions, respectively. $Li^+$ has a CN of four with both [TFSI] and [DCA], according to the number of coordination sites. However, the number of anions in the first solvation shell for [TFSI] is three, while it is four for [DCA]. This is a result of [TFSI] coordination *via* the oxygen atoms on the same anion where the two $Li^+$-O coordination *via* bidentate bonds with two [TFSI] anions yield a CN of four, as seen in

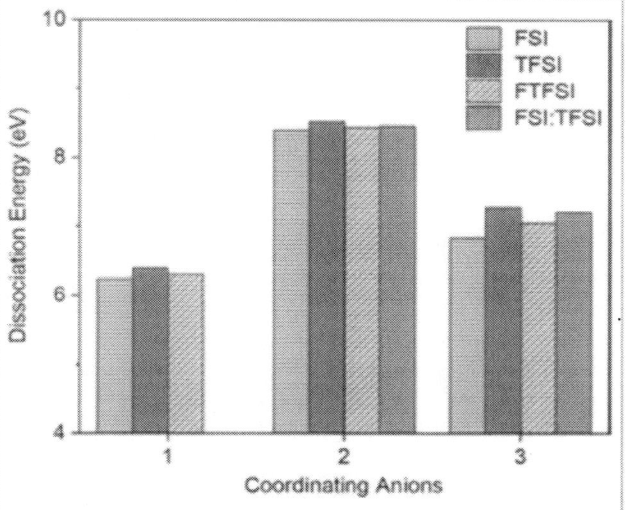


**FIGURE 10 |** Integral dissociation energies, $E_D$, of Li-[FSTFI] solvates in comparison to Li-[TFSI], Li-[FSI], and Li-[FSI]/[TFSI] mixture.

**Figure 5**. Previously, the integral binding energy for Li solvates with two [TFSI] anions calculated at the DFT/6-311+(d) level of theory was reported as −792.5 kJ/mol (8.2 eV) (Umebayashi et al., 2007), which is in reasonable agreement with our calculated Li $[TFSI]_2^{-1}$ (middle structure in **Figure 5**) $E_D$ of 8.59 eV (828.8 kJ/mol). For [DCA], there is $Li^+$-N interaction *via* monodentate bonds; therefore, four [DCA] anions yield a CN of four, as seen in **Figure 6**. Based on these results, $Li^+$-N coordination is stronger than $Li^+$-O, especially when solvates with 2–3 anions are compared, which is consistent with previously reported results (Huang et al., 2019b). Therefore, one can suggest that Li solvates with a [DCA] anion will be more rigid, and this can hinder $Li^+$ hopping in comparison to Li solvates with [TFSI]. This is counterintuitive to the low viscosities that are promoted with [DCA] anion in ILs. It should be recalled that while smaller and more rigid Li solvates may have higher vehicular motion, the hopping mechanism can prevail at high Li salt concentrations with weaker solvate structures.

To further understand the relationship of how the differing anions affect the strength of the interaction between $Li^+$ and the [TFSI] and [DCA] anions, the $E_D$ of the mixture system, [TFSI]/[DCA], as well as the individual anions are compared, as in **Figure 7**.

It should be noted that the minimum number of solvating anions for the [DCA]/[TFSI] system studied is two since a single anion would not capture the mixture behavior. Additionally, $Li^+$ solvation by three anions can either contain two [DCA] and one [TFSI] or *vice versa*; however, $Li^+$ coordination with two [DCA] anions provided a more thermodynamically favored structure (as shown in **Figure 7**). Comparing the $Li^+$ solvation energies of the [DCA]/[TFSI] mixture with those of [DCA]-only and [TFSI]-only systems, it is seen that with a CN of three, the mix-anion system (represented by orange bars in **Figure 7**) has a weaker $E_D$ than Li-[DCA] (blue bars) and a stronger $E_D$ than Li-[TFSI] (black) systems. For a perspective on how the mix-anion system of [DCA]/[TFSI] compares with the asymmetric analog of [CTFSI], the calculated dissociation energies are shown in **Figure 8**.

Based on the lower dissociation energies of Li solvates with the [CTFSI] anion, it can be said that the solvation cage is weakened by the anion asymmetry and the presence of competing coordination sites for $Li^+$. For example, comparing systems with two total solvation anions around $Li^+$, $(Li[CTFSI]_2)^{-1}$ has the most facile $E_D$ of 8.26 eV in comparison to 8.59 eV for $(Li[TFSI]_2)^{-1}$, 8.62 eV for $(Li[DCA][TFSI])^{-1}$, and 8.72 eV for $(Li[DCA]_2)^{-1}$. When [CTFSI] was compared with a slightly different asymmetric anion [CFSI], similar results were obtained, where Li-[CFSI] has more facile $E_D$ than the solvates with [FSI] or [DCA] alone. Further, Li-[CFSI] solvates had slightly reduced $E_D$ when compared to [CTFSI] (see **Supplementary Figure S4**). To understand the influence of solution composition on the $Li^+$ solvate, we performed an *ab initio* thermodynamic analysis of the mix-anion [DCA]/[TFSI] system. The thermodynamic stability for the anionic [DCA]/[TFSI] system as a function of [DCA] and [TFSI] mole fraction is presented in **Figure 9**. Indeed, these results indicate that $(Li[DCA]_2)^{-1}$ is thermodynamically most preferred except for very low mole fractions of [DCA], where $(Li[TFSI]_2)^{-1}$ is preferred. None of the mix-anion structures are thermodynamically preferred according to the *ab initio* thermodynamic analysis.

Taken together, the above analysis suggests that assestment of ion pair lifetimes and measurement of the $Li^+$ transport when solvated by asymmetric anions would be interesting and necessary to understand the transport mechanism as these

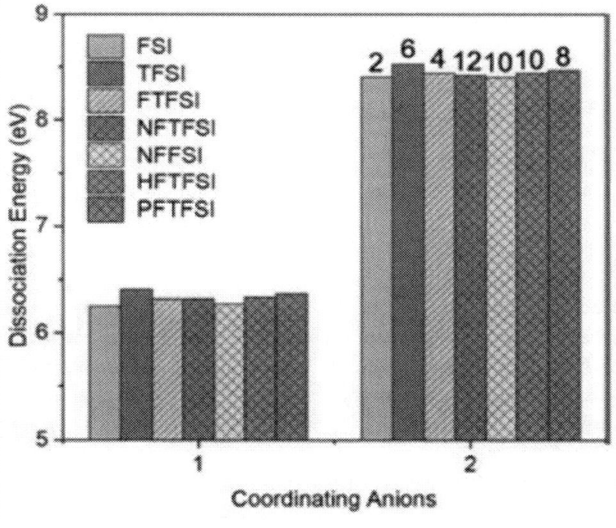


**FIGURE 11 |** Integral dissociation energies with respect to the number of coordinating asymmetric anions. [FTFSI] anion is highlighted with slashes, and all other asymmetric fluorinated anions are highlighted with crosshatches. PF, NF, and HF nomenclatures in front of TFSI or FSI stand for $C_2F_5$, $C_3F_7$, and $C_4F_9$ chains, respectively. The numbers on top of the bars indicate the number of total fluorines in the anion.

studies may provide a method to tune $Li^+$ hopping in concentrated electrolytes.

## $Li^+$ solvation by [FSI], [TFSI], and [FTFSI]

Li solvates with the asymmetric anion [FTSFI] were also studied in comparison to [TFSI], [FSI], and the [FSI]/[TFSI] mixture. It has been shown that $Li^+$ transport is enhanced in ILs with [FSI]/[TFSI] mix-anions, resulting in improved battery cycling over the Li-IL binary mixtures with either the [FSI] or the [TFSI] anion (Lux et al., 2009; Bayley et al., 2011; Huang et al., 2019a). In order to understand whether the improvements in LIBs with these electrolytes originated from a change in solvation energies, we calculated the dissociation energies for this system, similar to the analysis performed for $Li^+$-[CTFSI]. **Figure 10** shows that the asymmetric [FTFSI] behaves similarly to [FSI] and the [FSI]/[TFSI] mixture. It should be noted that the Li solvate with the [FSI]/[TFSI] mixture and three coordinating anions consists of two [TFSI] and one [FSI], as this produced the more thermodynamically favored structure. The other configuration with two [FSI] and one [TFSI] system resulted in a dissociation energy of 6.907 eV, which is slightly lower than coordination with three [FTFSI] anions for $(Li^+\text{-[FTFSI]}_3)^{-2}$ at 7.045 eV. Unlike the [CTFSI] asymmetric anion, Li solvates with [FTFSI] have higher $E_D$ than that of its parent anion [FSI], and the $E_D$ of all configurations is similar. This is due to the single coordination chemistry through the sulfonyl oxygen in all of the cases considered in **Figure 10**. The most significant difference among Li solvates with [TFSI], [FSI], and [FTFSI] is that with three coordinating anions, [FSI] has all monodentate $Li^+$-O(=S) coordination, leading to a CN of three, while [FTFSI] and [TFSI] have one bidentate and two monodentate bonds, respectively, leading to a CN of four, as seen in **Supplementary Figure S5**. All other structures demonstrated bidentate configurations where CN was two for a single solvating anion and four for two solvating anions. Due to the similarities in the solvation structure, the only impact on the solvation strength comes from the fluorinated groups, either F-S ([FSI]) or $CF_3$-S ([TFSI] and [FTFSI]).

## Fluorination Effects

The impacts of the length of fluorination and the degree of asymmetry on the calculated dissociation energies for Li solvates are captured in **Figure 11**. The asymmetry of the anions represented by the patterned bars in **Figure 11** is dictated by the length of the fluorinated alkyl chain on either the [FSI] or the [TFSI] parent anion. Specifically, the PF, NF, and HF nomenclatures in front of TFSI stand for $C_2F_5$, $C_3F_7$, and $C_4F_9$ chains, respectively, on the one side of the derivatized [TFSI] anion. As seen, there are no significant differences in $E_D$. The lowest dissociation energy is with [NFFSI] which has the longest fluorinated alkyl chains ($C_4F_9$) on the one side, with the other side being F-S.

While these results cannot point to a specific advantage of the longer fluorinated anions since they have the same enthalpic effects on $Li^+$ coordination, we expect to find entropic effects in solution which are not present in gas phase energy evaluations. In the case of [NFTFSI] and [NFFSI], the long fluorinated butane chains of each anion align with each other when the Li solvate has two anions with a CN of four *via* two bidentate bonds (**Supplementary Figure S6**). This alignment of the fluorinated alkyl chains may have a significant impact on the bulk liquid structure that cannot be evaluated by gas phase DFT calculations. We expect that the increase in the fluorination length will improve the viscosity and the solvation dynamics more than the enthalpic contributions to the dissociation energies, due to the delocalization of charge along the alkyl chain and inability of fluorochains solvating $Li^+$. In particular, the treatment of the entire system with IL cation and other cosolvents, if present, can provide more realistic solvate structures when longer chain fluorinated asymmetric anions are solvating $Li^+$. By studying these systems with MD simulations and experimental spectroscopic studies (Raman and NMR), the impact of fluorination on the bulk solvation and transport properties can be probed to a greater extent.

# CONCLUSION

The solvation structure and energies of $Li^+$ with solvating anions of [TFSI], [DAC], [CTFSI], and [FTFSI] and a number of longer fluorinated anions with $Li^+$-O and $Li^+$-N coordination capability were calculated. The anions studied produced Li solvate structures with CNs of 2–4 depending on the size and functionality of the anion. Li solvates with a CN of two resulted in the highest dissociation energies (most stable conformers). Li solvates with a CN of one resulted in the lowest $E_D$ with the exception of [DCA] and [CTFSI]. [DCA] and [CTFSI] were able to coordinate with four anions, and these Li solvates produced the lowest $E_D$ (weakest). Through the $E_D$ analysis, we found that the asymmetric [CTFSI] anion coordination with $Li^+$ is weaker than [TFSI] and [DCA] anions as well as the [TFSI]/[DCA] mixture. There are a larger number of solvate structures for Li-[CTFSI] solvate than those for [TFSI] and [DCA]. We suspect that this heterogeneity in the first solvation shell may lead to increased fluidity and structural diffusion (hopping) of $Li^+$ compared to solvates with [TFSI] and [DCA]. The [FTFSI] anion showed very similar results in terms of $E_D$ as the [TFSI] and [FSI] parent anions due to the similar coordination environments between the three anions. However, [FTFSI] yielded slightly lower $E_D$ than that of the $Li^+$ solvate with the parent anion mixture of [TFSI]/[FSI]. Given the weakened solvation energies, it would be interesting to study the impact of IL electrolytes with asymmetric anion on the $Li^+$ transference and performance of LIBs. The impact of fluorination length seems to have limited impact on $E_D$, with the longer fluorinated chains having slightly lower $E_D$ than [TFSI]. However, we anticipate that these long fluorinated alkyl chains will have more significant entropic effects, which are not captured in gas phase DFT calculations. By studying these systems with MD simulations and experimental spectroscopic studies (Raman and NMR), the impact of fluorination on the bulk solvation and transport properties can be probed to a greater extent. It will also be crucial for future studies to evaluate the impact of IL cations as well as the electric field present in Li batteries on the solvation structure and dynamics of $Li^+$.

## DATA AVAILABILITY STATEMENT

The original contributions presented in the study are included in the article/**Supplementary Material**; further inquiries can be directed to the corresponding author.

## AUTHOR CONTRIBUTIONS

DP performed the DFT calculations for geometries, vibrations, and dissociation energies. SV performed the *ab initio* thermodynamic analysis. RG supervised the *ab initio* thermodynamic analysis and contributed to the discussions. BG developed the plan, supervised DFT calculations, and contributed to the discussions. All authors contributed to the writing of the manuscript.

## FUNDING

BG would like to acknowledge funding from the National Science Foundation (CBET, Award No. 1903259). BG and DP would also like to acknowledge the High Performance Computing Resource in the Core Facility for Advanced Research Computing at Case Western Reserve University. SV is supported by a GAANN Fellowship (Award Number: P200A180076) from the United States Department of Education.

## SUPPLEMENTARY MATERIAL

The Supplementary Material for this article can be found online at: https://www.frontiersin.org/articles/10.3389/fenrg.2021.725010/full#supplementary-material

## REFERENCES

Armand, M., Endres, F., MacFarlane, D. R., Ohno, H., and Scrosati, B. (2009). Ionic-Liquid Materials for the Electrochemical Challenges of the Future. *Nat. Mater.* 8 (8), 621–629. doi:10.1038/nmat2448

Bae, S.-Y., Shim, E.-G., and Kim, D.-W. (2013). Effect of Ionic Liquid as a Flame-Retarding Additive on the Cycling Performance and thermal Stability of Lithium-Ion Batteries. *J. Power Sourc.* 244, 266–271. doi:10.1016/j.jpowsour.2013.01.100

Bauschlicher, C. W. (2018). Li+-Ligand Binding Energies and the Effect of Ligand Fluorination on the Binding Energies. *Chem. Phys. Lett.* 694, 86–92. doi:10.1016/j.cplett.2018.01.047

Bayley, P. M., Best, A. S., MacFarlane, D. R., and Forsyth, M. (2011). Transport Properties and Phase Behaviour in Binary and Ternary Ionic Liquid Electrolyte Systems of Interest in Lithium Batteries. *ChemPhysChem* 12 (4), 823–827. doi:10.1002/cphc.201000909

Borgel, V., Markevich, E., Aurbach, D., Semrau, G., and Schmidt, M. (2009). On the Application of Ionic Liquids for Rechargeable Li Batteries: High Voltage Systems. *J. Power Sourc.* 189 (1), 331–336. doi:10.1016/j.jpowsour.2008.08.099

Borodin, O., Giffin, G. A., Moretti, A., Haskins, J. B., Lawson, J. W., Henderson, W. A., et al. (2018). Insights into the Structure and Transport of the Lithium, Sodium, Magnesium, and Zinc Bis(trifluoromethansulfonyl)imide Salts in Ionic Liquids. *J. Phys. Chem. C.* 122 (35), 20108–20121. doi:10.1021/acs.jpcc.8b05573

Borodin, O., Smith, G. D., and Henderson, W. (2006). Li+ Cation Environment, Transport, and Mechanical Properties of the LiTFSI Doped N-Methyl-N-alkylpyrrolidinium+TFSI- Ionic Liquids. *J. Phys. Chem. B.* 110 (34), 16879–16886. doi:10.1021/jp061930t

Borodin, O., Suo, L., Gobet, M., Ren, X., Wang, F., Faraone, A., et al. (2017). Liquid Structure with Nano-Heterogeneity Promotes Cationic Transport in Concentrated Electrolytes. *ACS Nano* 11 (10), 10462–10471. doi:10.1021/acsnano.7b05664

Brinkkötter, M., Giffin, G. A., Moretti, A., Jeong, S., Passerini, S., and Schönhoff, M. (2018). Relevance of Ion Clusters for Li Transport at Elevated Salt Concentrations in [Pyr12O1][FTFSI] Ionic Liquid-Based Electrolytes. *Chem. Commun.* 54, 4278–4281. doi:10.1039/C8CC01416G

Castiglione, F., Ragg, E., Mele, A., Appetecchi, G. B., Montanino, M., and Passerini, S. (2011). Molecular Environment and Enhanced Diffusivity of Li+ Ions in Lithium-Salt-Doped Ionic Liquid Electrolytes. *J. Phys. Chem. Lett.* 2 (3), 153–157. doi:10.1021/jz101516c

Chen, F., and Forsyth, M. (2019). Computational Investigation of Mixed Anion Effect on Lithium Coordination and Transport in Salt Concentrated Ionic Liquid Electrolytes. *J. Phys. Chem. Lett.* 10 (23), 7414–7420. doi:10.1021/acs.jpclett.9b02416

Ditchfield, R., Hehre, W. J., and Pople, J. A. (1971). Self-Consistent Molecular-Orbital Methods. IX. An Extended Gaussian-Type Basis for Molecular-Orbital Studies of Organic Molecules. *J. Chem. Phys.* 54 (2), 724. doi:10.1063/1.1674902

Duluard, S., Grondin, J., Bruneel, J.-L., Pianet, I., Grélard, A., Campet, G., et al. (2008). Lithium Solvation and Diffusion in the 1-Butyl-3-Methylimidazolium Bis(trifluoromethanesulfonyl)imide Ionic Liquid. *J. Raman Spectrosc.* 39 (5), 627–632. doi:10.1002/jrs.1896

Eftekhari, A., Liu, Y., and Chen, P. (2016). Different Roles of Ionic Liquids in Lithium Batteries. *J. Power Sourc.* 334, 221–239. doi:10.1016/j.jpowsour.2016.10.025

Elmore, C., Seidler, M., Ford, H., Merrill, L., Upadhyay, S., Schneider, W., et al. (2018). Ion Transport in Solvent-free, Crosslinked, Single-Ion Conducting Polymer Electrolytes for Post-Lithium Ion Batteries. *Batteries* 4 (2), 28. doi:10.3390/batteries4020028

Frisch, M. J., Trucks, G. W., Schlegel, H. B., Scuseria, G. E., Robb, M. A., Cheeseman, J. R., et al. (2016). *Gaussian 16 Rev. C.01*. Wallingford, CT.

Fujii, K., Hamano, H., Doi, H., Song, X., Tsuzuki, S., Hayamizu, K., et al. (2013). Unusual Li+Ion Solvation Structure in Bis(fluorosulfonyl)amide Based Ionic Liquid. *J. Phys. Chem. C.* 117 (38), 19314–19324. doi:10.1021/jp4053264

Fulfer, K. D., and Kuroda, D. G. (2016). Solvation Structure and Dynamics of the Lithium Ion in Organic Carbonate-Based Electrolytes: A Time-dependent Infrared Spectroscopy Study. *J. Phys. Chem. C.* 120 (42), 24011–24022. doi:10.1021/acs.jpcc.6b08607

Galiński, M., Lewandowski, A., and Stępniak, I. (2006). Ionic Liquids as Electrolytes. *Electrochimica Acta.* 51 (26), 5567–5580. doi:10.1016/j.electacta.2006.03.016

Garcia, B., Lavallée, S., Perron, G., Michot, C., and Armand, M. (2004). Room Temperature Molten Salts as Lithium Battery Electrolyte. *Electrochim. Acta.* 49 (26), 4583–4588. doi:10.1016/j.electacta.2004.04.041

Getman, R. B., Xu, Y., and Schneider, W. F. (2008). Thermodynamics of Environment-dependent Oxygen Chemisorption on Pt(111). *J. Phys. Chem. C.* 112 (26), 9559–9572. doi:10.1021/jp800905a

Grundner, S., Markovits, M. A. C., Li, G., Tromp, M., Pidko, E. A., Hensen, E. J. M., et al. (2015). Single-site Trinuclear Copper Oxygen Clusters in Mordenite for Selective Conversion of Methane to Methanol. *Nat. Commun.* 6 (1), 7546. doi:10.1038/ncomms8546

Han, H., Guo, J., Zhang, D., Feng, S., Feng, W., Nie, J., et al. (2011). Lithium (Fluorosulfonyl)(nonafluorobutanesulfonyl)imide (LiFNFSI) as Conducting Salt to Improve the High-Temperature Resilience of Lithium-Ion Cells. *Electrochem. Commun.* 13 (3), 265–268. doi:10.1016/j.elecom.2010.12.030

Haskins, J. B., Bennett, W. R., Wu, J. J., Hernández, D. M., Borodin, O., Monk, J. D., et al. (2014). Computational and Experimental Investigation of Li-Doped Ionic

Liquid Electrolytes: [pyr14][TFSI], [pyr13][FSI], and [EMIM][BF4]. *J. Phys. Chem. B.* 118 (38), 11295–11309. doi:10.1021/jp5061705

Hoffknecht, J.-P., Drews, M., He, X., and Paillard, E. (2017). Investigation of the N -Butyl- N -Methyl Pyrrolidinium Trifluoromethanesulfonyl- N -cyanoamide (PYR 14 TFSAM) Ionic Liquid as Electrolyte for Li-Ion Battery. *Electrochim. Acta.* 250, 25–34. doi:10.1016/j.electacta.2017.08.020

Huang, Q., Lee, Y.-Y., and Gurkan, B. (2019a). Pyrrolidinium Ionic Liquid Electrolyte with Bis(trifluoromethylsulfonyl)imide and Bis(fluorosulfonyl) imide Anions: Lithium Solvation and Mobility, and Performance in Lithium Metal-Lithium Iron Phosphate Batteries. *Ind. Eng. Chem. Res.* 58, 22587–22597. doi:10.1021/acs.iecr.9b03202

Huang, Q., Lourenço, T. C., Costa, L. T., Zhang, Y., Maginn, E. J., and Gurkan, B. (2019b). Solvation Structure and Dynamics of Li+ in Ternary Ionic Liquid-Lithium Salt Electrolytes. *J. Phys. Chem. B.* 123 (2), 516–527. doi:10.1021/acs.jpcb.8b08859

Imai, M., Tanabe, I., Ikehata, A., Ozaki, Y., and Fukui, K.-i. (2020). Attenuated Total Reflectance Far-Ultraviolet and Deep-Ultraviolet Spectroscopy Analysis of the Electronic Structure of a Dicyanamide-Based Ionic Liquid with Li+. *Phys. Chem. Chem. Phys.* 22 (38), 21768–21775. doi:10.1039/D0CP03865B

Kerner, M., Plylahan, N., Scheers, J., and Johansson, P. (2015). Ionic Liquid Based Lithium Battery Electrolytes: Fundamental Benefits of Utilising Both TFSI and FSI Anions?. *Phys. Chem. Chem. Phys.* 17 (29), 19569–19581. doi:10.1039/C5CP01891A

Krachkovskiy, S., Dontigny, M., Rochon, S., Kim, C., Trudeau, M. L., and Zaghib, K. (2020). Determination of Binary Diffusivities in Concentrated Lithium Battery Electrolytes via NMR and Conductivity Measurements. *J. Phys. Chem. C.* 124 (45), 24624–24630. doi:10.1021/acs.jpcc.0c07383

Lassègues, J.-C., Grondin, J., Aupetit, C., and Johansson, P. (2009). Spectroscopic Identification of the Lithium Ion Transporting Species in LiTFSI-Doped Ionic Liquids. *J. Phys. Chem. A.* 113 (1), 305–314. doi:10.1021/jp806124w

Lee, C., Yang, W., and Parr, R. G. (1988). Development of the Colle-Salvetti Correlation-Energy Formula into a Functional of the Electron Density. *Phys. Rev. B.* 37 (2), 785–789. doi:10.1103/physrevb.37.785

Lesch, V., Jeremias, S., Moretti, A., Passerini, S., Heuer, A., and Borodin, O. (2014). A Combined Theoretical and Experimental Study of the Influence of Different Anion Ratios on Lithium Ion Dynamics in Ionic Liquids. *J. Phys. Chem. B.* 118 (26), 7367–7375. doi:10.1021/jp501075g

Lesch, V., Li, Z., Bedrov, D., Borodin, O., and Heuer, A. (2016). The Influence of Cations on Lithium Ion Coordination and Transport in Ionic Liquid Electrolytes: a MD Simulation Study. *Phys. Chem. Chem. Phys.* 18 (1), 382–392. doi:10.1039/C5CP05111H

Li, Z., Smith, G. D., and Bedrov, D. (2012). Li+ Solvation and Transport Properties in Ionic Liquid/Lithium Salt Mixtures: A Molecular Dynamics Simulation Study. *J. Phys. Chem. B.* 116 (42), 12801–12809. doi:10.1021/jp3052246

Lux, S. F., Schmuck, M., Appetecchi, G. B., Passerini, S., Winter, M., and Balducci, A. (2009). Lithium Insertion in Graphite from Ternary Ionic Liquid-Lithium Salt Electrolytes: II. Evaluation of Specific Capacity and Cycling Efficiency and Stability at Room Temperature. *J. Power Sourc.* 192 (2), 606–611. doi:10.1016/j.jpowsour.2009.02.066

Matsuda, Y., Fukushima, T., Hashimoto, H., and Arakawa, R. (2002). Solvation of Lithium Ions in Mixed Organic Electrolyte Solutions by Electrospray Ionization Mass Spectroscopy. *J. Electrochem. Soc.* 149 (8), A1045. doi:10.1149/1.1489687

McEldrew, M., Goodwin, Z. A. H., Bi, S., Kornyshev, A. A., and Bazant, M. Z. (2021). Ion Clusters and Networks in Water-In-Salt Electrolytes. *J. Electrochem. Soc.* 168 (5), 050514. doi:10.1149/1945-7111/abf975

Moučka, F., Nezbeda, I., and Smith, W. R. (2015). Chemical Potentials, Activity Coefficients, and Solubility in Aqueous NaCl Solutions: Prediction by Polarizable Force Fields. *J. Chem. Theor. Comput.* 11 (4), 1756–1764. doi:10.1021/acs.jctc.5b00018

Navarra, M. A. (2013). Ionic Liquids as Safe Electrolyte Components for Li-Metal and Li-Ion Batteries. *MRS Bull.* 38 (7), 548–553. doi:10.1557/mrs.2013.152

Noda, A., Hayamizu, K., and Watanabe, M. (2001). Pulsed-Gradient Spin−Echo1H and19F NMR Ionic Diffusion Coefficient, Viscosity, and Ionic Conductivity of Non-chloroaluminate Room-Temperature Ionic Liquids. *J. Phys. Chem. B.* 105 (20), 4603–4610. doi:10.1021/jp004132q

Nürnberg, P., Lozinskaya, E. I., Shaplov, A. S., and Schönhoff, M. (2020). Li Coordination of a Novel Asymmetric Anion in Ionic Liquid-In-Li Salt Electrolytes. *J. Phys. Chem. B.* 124 (5), 861–870. doi:10.1021/acs.jpcb.9b11051

Palafox, M. A. (2018). DFT Computations on Vibrational Spectra: Scaling Procedures to Improve the Wavenumbers. *Phys. Sci. Rev.* 3 (6), 1–30. doi:10.1515/psr-2017-0184

Palumbo, O., Trequattrini, F., Appetecchi, G., and Paolone, A. (2017). A Study of the Conformers of the (Nonafluorobutanesulfonyl)imide Ion by Means of Infrared Spectroscopy and Density Functional Theory (DFT) Calculations. *Challenges* 8 (1), 7. doi:10.3390/challe8010007

Paolucci, C., Parekh, A. A., Khurana, I., Di Iorio, J. R., Li, H., Albarracin Caballero, J. D., et al. (2016). Catalysis in a Cage: Condition-dependent Speciation and Dynamics of Exchanged Cu Cations in SSZ-13 Zeolites. *J. Am. Chem. Soc.* 138 (18), 6028–6048. doi:10.1021/jacs.6b02651

Pham, T. D., Bin Faheem, A., Chun, S. Y., Rho, J. R., Kwak, K., and Lee, K. K. (2021). Synergistic Effects on Lithium Metal Batteries by Preferential Ionic Interactions in Concentrated Bisalt Electrolytes. *Adv. Energ. Mater.* 11 (11), 2003520. doi:10.1002/aenm.202003520

Sakaebe, H., and Matsumoto, H. (2003). N-Methyl-N-propylpiperidinium Bis(trifluoromethanesulfonyl)imide (PP13-TFSI) - Novel Electrolyte Base for Li Battery. *Electrochem. Commun.* 5 (7), 594–598. doi:10.1016/s1388-2481(03)00137-1

Shen, S., Fang, S., Qu, L., Luo, D., Yang, L., and Hirano, S.-i. (2015). Low-viscosity Ether-Functionalized Pyrazolium Ionic Liquids Based on Dicyanamide Anions: Properties and Application as Electrolytes for Lithium Metal Batteries. *RSC Adv.* 5 (114), 93888–93899. doi:10.1039/C5RA17539A

Shim, Y. (2018). Computer Simulation Study of the Solvation of Lithium Ions in Ternary Mixed Carbonate Electrolytes: Free Energetics, Dynamics, and Ion Transport. *Phys. Chem. Chem. Phys.* 20 (45), 28649–28657. doi:10.1039/C8CP05190A

Soon, A., Todorova, M., Delley, B., and Stampfl, C. (2007). Surface Oxides of the Oxygen-Copper System: Precursors to the Bulk Oxide Phase?. *Surf. Sci.* 601 (24), 5809–5813. doi:10.1016/j.susc.2007.06.062

Starzak, M. E. (2010). *Energy and Entropy.* New York, NY: Springer-Verlag.

Sun, Y., Xin, N., and Prausnitz, J. M. (2018). Solubilities of Five Lithium Salts in 1-Butyl-3-Methylimidazolium Dicyanamide and in 1-Butyl-3-Methylimidazolium Tetrafluoroborate from 298.15 to 343.15 K. *J. Chem. Eng. Data.* 63 (12), 4524–4531. doi:10.1021/acs.jced.8b00618

Umebayashi, Y., Mitsugi, T., Fukuda, S., Fujimori, T., Fujii, K., Kanzaki, R., et al. (2007). Lithium Ion Solvation in Room-Temperature Ionic Liquids Involving Bis(trifluoromethanesulfonyl) Imide Anion Studied by Raman Spectroscopy and DFT Calculations. *J. Phys. Chem. B.* 111 (45), 13028–13032. doi:10.1021/jp076869m

Umebayashi, Y., Mori, S., Fujii, K., Tsuzuki, S., Seki, S., Hayamizu, K., et al. (2010). Raman Spectroscopic Studies and *Ab Initio* Calculations on Conformational Isomerism of 1-Butyl-3-Methylimidazolium Bis-(trifluoromethanesulfonyl) amide Solvated to a Lithium Ion in Ionic Liquids: Effects of the Second Solvation Sphere of the Lithium Ion. *J. Phys. Chem. B.* 114 (19), 6513–6521. doi:10.1021/jp100898h

Yamada, Y., Wang, J., Ko, S., Watanabe, E., and Yamada, A. (2019). Advances and Issues in Developing Salt-Concentrated Battery Electrolytes. *Nat. Energ.* 4 (4), 269–280. doi:10.1038/s41560-019-0336-z

Yang, L., Xiao, A., and Lucht, B. L. (2010). Investigation of Solvation in Lithium Ion Battery Electrolytes by NMR Spectroscopy. *J. Mol. Liquids.* 154 (2), 131–133. doi:10.1016/j.molliq.2010.04.025

Yoon, H., Lane, G. H., Shekibi, Y., Howlett, P. C., Forsyth, M., Best, A. S., et al. (2013). Lithium Electrochemistry and Cycling Behaviour of Ionic Liquids Using Cyano Based Anions. *Energy Environ. Sci.* 6 (3), 979–986. doi:10.1039/C3EE23753B

**Conflict of Interest:** The authors declare that the research was conducted in the absence of any commercial or financial relationships that could be construed as a potential conflict of interest.

**Publisher's Note:** All claims expressed in this article are solely those of the authors and do not necessarily represent those of their affiliated organizations, or those of the publisher, the editors, and the reviewers. Any product that may be evaluated in this article, or claim that may be made by its manufacturer, is not guaranteed or endorsed by the publisher.

# 625496 Organic Solvent Membrane Filtration: Rational Design of Bottlebrush Membranes

**Abstract:**

Our long-term goal is to fractionate crude-oil mixtures and challenge phase-change distillation with low energy non-phase change membrane filtration using graft chemistry and transport phenomena. We have developed and tested a new series of bottle-brush filtration membranes that have flexibility in chemistry, structure and morphology (these properties are not available in current commercial membranes) and are stable in organic solvents. So far we have tested them with a series of alcohols and the dye Rose Bengal (Ramesh et al. (2021) "Organic solvent filtration by brush membranes: Permeability, selectivity and fouling correlate with degree of SET-LRP grafting", *J. Membr. Sci.618, 118699)*. The permeability of these composite membranes, evaluated with 6 organic solvents (alcohols), correlates with solvent properties and can be tailored with degree of grafting (DG). Filtration performance was assessed by measuring the selectivity and fouling of the organic dye Rose Bengal (RB) in ethanol. The polarized wall concentration increased with DG, whereas the RB sieving coefficient decreased with DG. Flux decline during RB filtration was best described by pore blocking and by heterogeneous and Langmuir sorption models. These studies provide a basis for designing polymeric membranes comprising branched brush selective layers with defined microstructure not possible with the widely popular phase inversion process. Our latest unpublished results demonstrate that these membranes can be tuned to distinguish between n-butanol and iso-butanol and between iso-propanol and n-butanol, so we are hopeful that they will also be able to fractionate light crude oil and eventually crude-oil.

# 625545 Cell-Free Production of Isobutanol: A Completely Immobilized System

**Abstract:**

Our long-term goal is to synthesize fuels efficiently and economically using *in vitro* (i.e. cell free) technology by minimizing diffusion limitations (i.e. at high Damkohler number) with an immobilized multi-enzyme system. A completely immobilized cell-free enzyme reaction system was used to convert ketoisovaleric acid to iso- butanol, a desirable biofuel, with a molar yield of 43% and a titer of 2 g/L, which are comparable to high performing in vivo systems (e.g. 41% and 5.4 g/L, respectively, for Clostridium thermocellum). The approach utilizes, for the first time, a series of previously reported enzyme mutants that either overproduce the product or are more stable when compared with their wild type (Wong, M., Sorci, M., Gasparis, G., Belfort,G., Koffas, M. (2020) "Cell-Free Production of Isobutanol- A completely immobilized system" *Bioresource Technol.,* 294, 122104. ). The selected enzyme variants include keto-acid dec- arboxylase attached to a maltose binding protein, alcohol dehydrogenase, and formate dehydrogenase. These enzymes were screened for thermal, pH, and product stability to choose optima for this system which were pH 7.4 and 35 °C. This system is designed to address well-known limitations of in vivo systems such as low product concentrations due to product feedback inhibition, instability of cells, and lack of economic product recovery.

# 626700 Solid State Battery Based on PEO/PVDF-HFP Solid Polymer Electrolyte

**Abstract:**

In this work, a solid polymer electrolyte (SPE) was synthesized by complex of PEO, PVDF-HFP, and LiTFSI. The SPE has high thermal stability and can be used in high temperature working environment. Solid-state lithium batteries based on SPE, lithium metal, and $LiFePO_4$ were assembled. At 70 °C and 120 °C, the obtained lithium batteries demonstrated excellent electrochemical performance, indicating that the tailor-made SPE has a wide working temperature range and it is suitable for high-performance solid-state batteries. A symmetrical Li/SPE/Li battery operated at 120 °C for 1,300 h with low polarization voltage and without any short circuit, indicating that the interface compatibility between lithium metal and SPE film was good and could effectively suppress the formation of lithium dendrites. The superior performance of the resulting SPE was mainly due to high thermal stability of the SPE, the interaction between the lithium salt and the host polymer, as well as weak interaction between PEO and PVDF-HFP, which reduced the crystallinity of PEO, thereby significantly improving the $Li^+$ migration ability and improving the battery's cycle life.

# 626704 Ion Transport and Ion-Correlation in Non-Aqueous Lithium-Ion Polyelectrolyte Solutions

**Abstract:**

A majority of ionic conductivity in current state-of-the-art liquid electrolytes results from the movement of the anion rather than the electrochemically active $Li^+$ ion. Passing current through such an electrolyte can cause severe ionic concentration gradients to extend into porous electrodes, which at high currents can reduce active material utilization and result in high overpotentials that cause deleterious side reactions to occur (e.g., Li plating on graphite electrodes)[1]. The severity of these concentration gradients is directly influenced by the transference number ($t_+$), which describes the ratio of current carried by the electrochemically active $Li^+$ ion to the total current passed. Non-aqueous polyelectrolyte solutions (PESs) have been recently suggested as a promising route to high transference number electrolytes that demonstrate potential for both high ionic conductivity and $t_+$[2–4]. This combination is achieved by anchoring the anion to a polymer backbone – allowing for high $t_+$ by slowing down the motion of the electrochemically inactive anion – while maintaining high ion conductivity through improved ion dissociation and solvent-mediated $Li^+$ transport. Recent molecular dynamic simulations of PESs have highlighted the critical importance of correlated ion motion in these systems and have called into question oligomeric PESs as a feasible strategy to achieving high $t_+$ and conductivity electrolytes[5,6].

In this presentation we discuss our efforts to date synthesizing and experimentally studying ion transport in lithium-ion and lithium metal battery-relevant PESs- specifically lithium triflimide appended polystyrene (PS-LiTFSI) and polymethacrylate (PA-LiTFSI) dissolved in carbonate blends. Using a combination of pulsed-field gradient NMR, Raman spectroscopy, and electrochemical experiments, we characterized the transport properties, including the conductivity, diffusion coefficients, and $t_+$ of PS-LiTFSI and PA-LiTFSI PESs. We seek to understand the role of correlated ion motion in these systems and the impact that these correlations have on lithium-ion transport parameters relevant to battery performance. Comparing our experiments[7] to insights from molecular dynamics modeling we demonstrate that despite selectively slowing anion motion using polyanions, anion-anion correlation through the polymer backbone likely contributes to reducing the $t_+$ in PESs. We will discuss future strategies to tune this effect by manipulating polyion molecular weight, backbone chemistry, and charge distribution on the polymer chain, particularly in the entangled high-molecular weight regime.

<u>References</u>

(1) Diederichsen et al. *ACS Energy Lett.* **(2017)**
(2) Diederichsen et al. *Macromolecules* **(2018)**
(3) Diederichsen et al. *J. Phys. Chem. B* **(2019)**
(4) Dewing et al. *Chem. Mater.* **(2020)**
(5) Fong et al. *ACS Cent. Sci.* **(2019)**
(6) Fong et al. *Macromolecules* **(2021)**
(7) Bergstrom et al. *ECSarXiv* **(2021)**

# 622445 Studying the Mechanism of the High Voltage $Ce^{3+}/Ce^{4+}$ Redox Couple through Kinetic Measurements and Spectroscopy

**Abstract:**

Renewable energy is the fastest growing source of electricity in the U.S., and is predicted to continue to be the fastest growing source through at least 2050, based on current incentives and falling technology costs.[1] Due to the intermittent nature of renewable energy, advances in energy storage technologies are key to integrating renewable energy into the U.S. electricity grid. Redox flow batteries (RFBs) are a promising electrochemical technology for large scale energy storage, including grid-scale storage of renewable energy, due to their longer lifecycles and easier scalability and modularity than other, more developed battery technologies, e.g., lithium-ion.[2] Despite their promising nature, RFBs are currently too expensive for market deployment. For instance, the U.S. Department of Energy has reported a target capital cost for new energy storage technologies of 150 $/kWh,[3] yet recent reports of the all-vanadium RFB indicate the current capital cost is at least 555 $/kWh.[2] The large expense of RFBs is in part driven by inefficiencies in mass transport, ohmic, and kinetics, resulting in large losses in voltage at high current densities. One way to improve the efficiency of RFBs is to explore new chemistries that increase the voltage window of the battery, effectively reducing the significance of any incurred voltage losses as well as increasing the power density. The cerium redox reaction between $Ce^{3+}$ and $Ce^{4+}$ is a promising candidate for the active species at the positive electrode in an aqueous RFB due to its high positive redox potential that ranges between 1.28 V vs SHE in HCl and 1.74 V vs SHE in $HClO_4$.[4] Despite being thermodynamically promising for RFB applications, the kinetics of the $Ce^{3+}/Ce^{4+}$ electron transfer have been reported to be a limiting performance factor in studies of cerium RFB systems.[5] Any efforts to improve the rate of the cerium electron transfer are hindered by the dearth of clear information on the cerium charge transfer, i.e., there are conflicting reports on the nature of the redox mechanism as well as a lack of knowledge on the structure of the $Ce^{3+}$ and $Ce^{4+}$ ions. In this talk we will discuss the identification of the structures of the ionic species involved in the charge transfer as well as identify if $Ce^{3+}/Ce^{4+}$ is an inner-sphere or outer-sphere reaction. We propose a possible two-step outer-sphere electron transfer mechanism that satisfies both our observed reaction kinetic measurements in sulfuric acid and structural information obtained from X-ray absorption and UV-Vis spectroscopy.

Whether the cerium electron transfer occurs through an inner-sphere or outer-sphere mechanism will greatly influence the approach to improving the kinetics of the reaction. An inner-sphere heterogeneous electron transfer occurs when the reactant, intermediate, or product interacts strongly with the electrode surface. An outer-sphere reaction is one in which the coordination spheres of the reactant species are maintained in the activated complex, meaning the reactant and product do not interact strongly with the electrode surface.[6] If the $Ce^{3+}/Ce^{4+}$ electron transfer is inner-sphere, then it is expected that the electrode material will control the rate of the reaction, and future research efforts should be focused on isolating the key properties of electrodes that improve the cerium redox kinetics. If the $Ce^{3+}/Ce^{4+}$ electron transfer is outer-sphere, then the electrode material must simply be conductive and stable, and instead the electrolyte will control the rate of electron transfer through the reorganization energy. Thus, if it is determined that the cerium electron transfer is outer-sphere, research into electrolyte properties that affect the reorganization energy of cerium ions should be prioritized.

In this work, we will first present Extended X-Ray Absorption Fine Structure (EXAFS) evidence at the Ce K-edge that suggests an inner-sphere structural change occurs as $Ce^{3+}$ is oxidized to $Ce^{4+}$ in sulfuric acid. This study builds off of our previous work,[7] which used EXAFS at the Ce $L_3$-edge as well as density functional theory (DFT) calculations to show that $Ce^{3+}$ is hydrated by nine water molecules while $Ce^{4+}$ is complexed by at least one electrolyte anion in several aqueous solutions. Here, the use of the Ce K-edge ensures higher quality EXAFS data at higher k values than was possible at the $L_3$-edge, allowing us to probe the Ce ion structures at higher coordination shells. We collected EXAFS data for solutions of 0.05 M $Ce^{3+}$ in 2 M $H_2SO_4$ and 0.05 M $Ce^{4+}$ in 2 M $H_2SO_4$ using the transmission mode at the Ce K-edge. By fitting the experimental EXAFS spectra for both $Ce^{3+}$ and $Ce^{4+}$ with relevant scattering shells, e.g., Ce-O, Ce-H, Ce-S, we will report the coordination numbers and bond distances of the local environment for both $Ce^{3+}$ and $Ce^{4+}$ ions in sulfuric acid. We will also discuss our comparison between the experimentally collected EXAFS and Molecular Dynamics EXAFS (MD-EXAFS) of several possible $Ce^{3+}$ and $Ce^{4+}$ structures ($[Ce(H_2O)_9]^{3+/4+}$ or $[Ce(H_2O)_8SO_4]^{+/2+}$ complexes). The structures and surrounding water were generated using Carr-Parrinello molecular dynamics in the NWChem software, which were then used as inputs to the FEFF9 code to calculate EXAFS signals. The MD-EXAFS will be used to corroborate our conclusion that an inner-sphere structural change occurs as the cerium electron transfer occurs.

Although there is an inner-sphere change from $Ce^{3+}$ to $Ce^{4+}$, our kinetic measurements imply that the observed rate follows an outer-sphere mechanism, i.e., it is independent of the electrode surface. We compare the kinetic activity of glassy carbon (GC) and platinum (Pt) rotating disk electrodes (RDEs) in sulfuric acid by extracting the exchange current densities for each electrode material as a function of state of charge (SoC). The exchange current densities at each SoC were collected from both Tafel plots and charge transfer resistance values in a solution of 2 M $H_2SO_4$ with 0.05 M total cerium. RDEs were used for both electrode materials to control for mass transfer effects by confirming that exchange current densities were independent of rotation rate. The exchange current densities as a function of SoC could be fitted well to the Butler-Volmer equation[6] for both the Pt and GC electrodes. By fitting the data for both the Pt and GC RDEs, we extracted an experimental rate constant and charge transfer coefficient for both electrodes. We determined that the rate constants for the $Ce^{3+}/Ce^{4+}$ electron transfer differed by a factor of 5.5 between the Pt and GC RDEs, which is a significantly lower factor than those observed for inner-sphere redox couples, and in line with the electrode dependence of well-known outer-sphere reactions (e.g., $[Fe(CN)_6]^{4-/3-}$, $[Ru(NH_3)_6]^{2+/3+}$). These findings suggests that the $Ce^{3+}/Ce^{4+}$ electron transfer occurs via an outer-sphere mechanism.

To reconcile the seeming contradiction between the outer-sphere kinetic behavior and the inner-sphere structural change, we propose a reaction mechanism in which the sulfate-complexed $Ce^{4+}$ is reduced to a high-energy sulfate-complexed $Ce^{3+}$ species through an outer-sphere electron transfer, which then undergoes a fast (quasi-equilibrated) ligand exchange with water to form the low energy $[Ce(H_2O)_9]^{3+}$, completing the charge transfer. Although the kinetic data fits well to the Butler-Volmer equation for a single step outer-sphere mechanism, this is not consistent with the inner-sphere structural change that we show occurs between $Ce^{3+}$ and $Ce^{4+}$ in $H_2SO_4$ through EXAFS. Thus, a more realistic way to fit the data would be to use Marcus Theory to describe the rate law for the rate-determining electron transfer step, while incorporating the quasi-equilibrated ligand exchange. We will show our experimental exchange current density data for both the Pt and GC RDEs as fit to a rate law derived from the proposed reaction mechanism to extract relevant parameters, including reorganization energy and activation barriers. We show that these experimentally derived values can be compared well to those obtained through DFT modeling predictions, which were based on the free energies of the $Ce^{3+}$ and $Ce^{4+}$ ions in sulfuric acid. The agreement demonstrates that we have identified a charge transfer mechanism that is consistent with both observed kinetic behavior and first principles and represents a significant advancement in the field of cerium redox kinetics. The knowledge that the cerium redox reaction displays outer-sphere behavior will advance RFB technologies by showing that future Ce-RFB work should focus on optimizing the properties of the electrolyte that maximize kinetic performance without sacrificing thermodynamic redox potentials. To parallel the necessary electrolyte studies, future Ce-RFB electrode research can thus prioritize stability in acidic solutions over inherent catalytic activity. In addition to this study's contribution to RFBs, it demonstrates a comprehensive way to couple structural information from spectroscopy with kinetic measurements as well as first principles calculations to test the validity of a hypothesized electron transfer mechanism, which is a technique that can be applied to proposed mechanisms for many relevant electrochemical reactions beyond cerium.

1. U.S. EIA. *Annual Energy Outlook 2021*. (2021).

2. PNNL. *Energy Storage Technology and Cost Characterization Report*. (2019).

3. U.S. DOE. *Energy Storage Program Planning Document*. (2011).

4. Smith, G. F. & Getz, C. A. *Ind. Eng. Chem. Res.* **10**, 191–195 (1938).

5. Walsh, F. C. *et al. Chempluschem* **80**, 288–311 (2015).

6. Bard, A. J. & Faulkner, L. R. (John Wiley & Sons, Inc., 2001).

7. Buchanan, C. A. *et al. Inorg. Chem.* **59**, 12552–12563 (2020).

# 632780 Thermal Conductivity of Imidazolium Ionic Liquids Saturated with Compressed Gases

**Abstract:**

The low volatility of ionic liquids (ILs) allows for their potential usage in many engineering applications. Such applications often include systems of gases compressed in ionic liquids. To properly design and characterize such systems, transport property data is needed. Here, the thermal conductivity of several imidazolium based ionic liquids saturated with various compressed gases is measured. Systems reported and analyzed in this study include imidazolium ionic liquids saturated with several hydrofluorocarbon (HFC) gases as well as carbon dioxide at temperatures ranging up to 125 ºC and pressures ranging up to 200 bar. For the various IL/HFC systems investigated herein, a slight decrease in thermal conductivity is observed with increasing gas composition. For $CO_2$/IL systems, a slight decrease in thermal conductivity is observed with increasing gas compositions followed by a gradual increase at higher compositions. The increase in thermal conductivity is attributed to hydrostatic effects as the $CO_2$ is soluble in the ILs over a broad pressure range. Nevertheless, in all systems investigated here, the thermal conductivity of the mixture remains dominated by that of the pure IL even at large compositions of dissolved gas. This implies that simple composition-based mixing rules for mixture properties based upon pure component properties would not be appropriate.

# 623105 Epoch-Making Breakthrough Electric Generator

**Abstract:**

Epoch-making breakthrough electric generator

Author: David Judbarovski, systems engineering, principle inventor,.retired

judbarovski@gmail.com

## Abstract

Invented heterodyne antenna thermo electric generator is very cheap and small and light and without any moving parts. Its energy efficiency is near 100%. Carnot limit is not relevant.

Power generation and hybrid electric vehicles would consume some times less fuels, so generating some times less exhausts, and I have developed a technology for very mass and cheap CO2 and H2 from the air and water for the fuels recycling.

## Disclosure for experts

(*for lighter understanding see a popular but clear introduction below*)

Antenna is jointed with radiators, heated by heat carrier or by any combustible or by solar beams concentrator. The said radiator emits a band of EM waves of about 100 THz interfered each with the others forming envelope band, can be filtered by the said antenna resonators of much less frequencies into electricity with high energy efficiency that can reach near 100% especially after recuperation of exhaust heat.

If using Plank's Law on-line calculator for blackbody, the Radiant emittance is W = 50.8 kW/m2, and if Band Radiance chosen to be 10 THz around maximum, at 700 Celsius the said efficiency is 92%. and 109 kW/m2, 76% at 900 C. CAPEX is an order of magnitude cheaper than secondary battery

## Popular introduction for lighter understanding

The novel understanding can be proved by a simplest experiment. If blue band and yellow band are falling mixed on white sheet, it creates a jointed band comprised two separate bands being green one and infrared band approximately from 1.5 um to 2.5 um.

Ideally, a heterodyne can serve for transforming a high frequency A by a mixing it with a frequency B = A – E, and E << A, into envelope of much lower frequency E carried by the said high frequency A. Then we can high energy effective extract and use the said low frequency E by filtering the said mix, e.g. by resonator of the frequency E or by a rectifier of frequency lower than A, but equal or bigger then E..

Really It is known from school, that

$$\text{Sin}(A*t + C) + \text{Sin}(B*t) = 2 * (\text{Sin}(t * (A + B)/2 + C/2) * \text{Cos}(t * (A - B)/2 + C/2)$$

If A is a band of frequencies growing continuously from the band's ends to inside maximal value, it can be considered as a sum of separate frequencies recurrently interfered each with the others, so if being fallen on the resonator equipped by the said rectifier, the result would be direct current.

Heat can be considered as such wide band electromagnetic radiance, analogously to solar radiation, and antenna receiver can be made with an array of resonators of a width of frequency band have to be enough to satisfy frequency of feasible rectifiers, and much lower than maximal frequency of the heat radiance.

# 626877 Modeling Hierarchical Li-Air Cathode Designs with Improved Discharge Capacity


**Abstract:**

The emergence of Li-air battery, particularly non-aqueous Li-air battery have drawn much attention as a promising power source for next-generation electric vehicles. However, prior to its viable commercialization, there are some technical obstacles that need to be resolved (1). Among them, lower discharge capacity is one of the main challenge which is induced by the generation of insoluble lithium peroxide ($Li_2O_2$), during discharging, inside the cathode pores. Therefore, optimization and designing novel cathodes have become critically acclaimed topic among the scientific community. Extensive experimental and modeling investigations (2-6) have been carried out to develop novel cathode structures. All these studies ended up with the development of functionalized carbon (CNT, CNF, and graphene etc.) cathodes, hierarchical cathode structures functionalized with electrocatalysts (2, 3), and cathode structures with varying wettability (4, 7, 8). These studies focused on enhancing the porosity and pore surface characteristics, resulting in significant improvement in specific discharge capacity. Recently, it is reported (9) that the stochastic nature of pores interconnectivity (called tortuosity) and composite cathodes morphologies may inevitably enhance capacity due to wide range of pore size distribution (micro-, meso-, and macro-pores) and large accessible surface area. Thus, it is deduced that the concept of distributed tortuosity and composite materials also need to be considered in the designing of novel hierarchical cathode structures.

In this presentation, we will present and discuss results concerning a detailed model of Li-Air battery that we have developed and validated in COMSOL. The model helps in examining the potential of hierarchical cathode in guiding and enhancing the efficiency of transport phenomena and discharge product formation inside micro-, meso-, and macro-pores. Discharge curves results of hierarchical cathodes with distributed tortuosity and porosity (higher tortuosity with lower porosity on separator side, and lower tortuosity with higher porosity on cathode airside) considering a discharge current of 0.1 milliampere per square centimeter ($mA/cm^2$) are shown in Figure 1. It is found that this hierarchal cathode have significantly enhanced the discharge capacity by facilitating the effective transport of oxygen and distributing Li ions along the pores which in turn avoid blocking the pore space. This implies that hierarchical cathode materials with distributed tortuosity and porosity improved the pore alignment, effective transport of oxygen, active reactive area, and space to accommodate the solid discharge product. Despite containing same average porosity (Figure 1b), different porosity distributions resulted in different specific discharge capacities because of different utilization rates of their corresponding pore systems. It is evident that the discharge capacity attained in exponential case (red line) is larger than that in uniform initial porosity (dark gray line). This is because the exponential distribution in porosity effectively facilitates the pore distribution, pore volume, and accessible surface area resulting in facile transport and maximum accommodation of $Li_2O_2$ compared to the rest of the cases.



We acknowledge financial support for this work from Khalifa University of Science and Technology (projects RC2-2019-007 and CIRA2018-103).


**References**


1. Z. Ma, X. Yuan, L. Li, Z. F. Ma, D. P. Wilkinson, L. Zhang and J. Zhang, Energy and Environmental Science, 8, 2144 (2015).
2. H. W. Junrong Shen, Wang Sun, Qibing Wu, Shuying Zhen, Zhenhua Wang, Kening Sun, Journal of Materials Chemistry A, 7, 10662 (2019).
3. S. Hyun, B. Son, H. Kim, J. Sanetuntikul and S. Shanmugam, Applied Catalysis B: Environmental, 263, 118283 (2020).
4. P. Tan, W. Shyy and T. Zhao, Science Bulletin, 60, 975 (2015).

5. K. Jiang, X. Liu, G. Lou, Z. Wen and L. Liu, Journal of Power Sources, 451, 227821 (2020).
6. F. Y. Jie Li, Zipeng Su, Tianyu Zhang, Xiaochen Zhang, Hong Sun, Journal of The Electrochemical Society, 167, 090529 (2020).
7. J. W. Jung, S. H. Cho, J. S. Nam and I. D. Kim, Energy Storage Materials, 24, 512 (2020).
8. P. Tan, H. R. Jiang, X. B. Zhu, L. An, C. Y. Jung, M. C. Wu, L. Shi, W. Shyy and T. S. Zhao, Applied Energy, 204, 780 (2017).
9. A. Torayev, A. Rucci, P. C. M. M. Magusin, A. Demortière, V. De Andrade, C. P. Grey, C. Merlet and A. A. Franco, Journal of Physical Chemistry Letters, 9, 791 (2018).

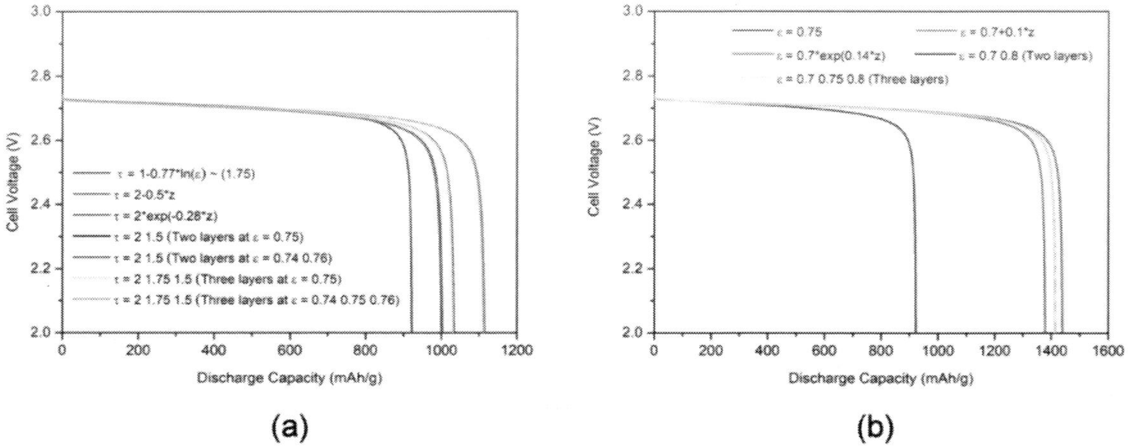


**Fig.1:** Schematics of discharge capacities of hierarchical cathodes with (a) distributed tortuosity, and (b) porosity- distributed porosity at 0.1 mA/cm$^2$.

# 629744 Light-Hydrocarbons Diffusion Coefficients in Semi-Crystalline Polyolefins: Automated Experiments with High Performance Differential Pressure-Decay

**Abstract:**

Transport of monomer(s) is one of essential topics in polymer reaction engineering. Understanding the diffusion of gaseous penetrants in complex semicrystalline polymer structure is still not complete despite of the benefits possibly leading to a better reactor control. The aim of this work is the investigation of diffusion in semicrystalline polyolefins by experimental and theoretical methods. Experiments were conducted in our "self-made" apparatus – differential pressure-decay.

Pressure decay is the method used to observe transport dynamics of gaseous species in solid material or their melts. The solid sample – the polymer in our case is places in the closed cell. For short time are connections to cell opened and the penetrating gas is injected there. When the connection into cell is closed again pressure evolution is observed and recorded. Pressure is decreasing due to sorption into a polymer sample. The equilibrium state is indicated by a stable pressure and the experimental procedure continues with fast cell degassing. The rising pressure evolution now represents the penetrant desorption process from polymer sample. These two main subprocesses are repeated periodically to obtain high number of experimental diffusion curves needed for the diffusion coefficients estimation procedure.

The apparatus is designed to hold up pressures occurring commonly in industry (up to 30 bar) and sufficient temperature range (30 °C – 110 °C). The overall system (e.g. solenoid valves, cells, connections, pressure transducers) are placed in the insulated box and are heated by hot air circulation to achieve uniform temperature.

Fast pressure changes during experimental procedure are associated with temperature fluctuations affecting the actual pressure in the investigated system. To reduce this phenomenon the pressure decay apparatus was recently rebuilt to add reference cell with high precious differential pressure transducer measuring pressure difference between cells. The reference cell is empty and the differential pressure evolution is not affected by temperature changes after the step change of pressure.

The assembled apparatus will serve for systematic and automated examination of diffusion coefficients in polymers. The control and data acquisition is realized by LabView script.

The objective of our research is to investigate the diffusion of light-hydrocarbon olefins such as ethylene, propylene or n-hexane in semi-crystalline polymers. Polyolefin morphology with stacked crystalline lamellae and the presence of two different amorphous regions is not commonly known and its scheme is shown in Figure 1 below. During the diffusion the penetrating molecules have to go through these amorphous regions having a different permeability. In this poster we will present diffusion coefficients of light-hydrocarbon olefins in various density polyethylenes (PE) and polypropylenes (PP). Density of polymer affects the physical properties and is directly connected to polymer crystallinity. We present estimated diffusion coefficients as a function of penetrant volume fraction in amorphous regions, as this allows a good comparison for various penetrants. Equilibrium sorption isotherms of penetrants in polyolefins are obtained by our magnetic-suspension gravimetric apparatus. The calculations must also include the crystallinity correction according to [1] at current temperature. In previous work [2] we have shown that for ethylene sorption isotherms the swelling phenomenon cannot be neglected and thus we must correct also geometric dimensions of evaluated polymer sample during diffusion coefficient computation.

Diffusion coefficient evaluation is accomplished with Fickian approach which is the core of the advanced evaluation program which is also reconstructing the beginning and end of the diffusion curve at the time where the solenoid valves are opened, and data cannot be collected. The only output parameter is then diffusion coefficient computed by Finite Volume Method.

The literature often proposes that the diffusion of light-hydrocarbons in polymers is concentration dependent. Fickian approach is not capable to describe this dependency by theoretically-supported model. The often employed approach is Free Volume Theory (FVT) [3] which supports the concentration dependent diffusion coefficient. The principle of this theory requires to estimate two fundamental constants of the polymer sample. By employing of FVT we shall be able to predict other penetrants diffusion coefficients in previously examined polymer samples.

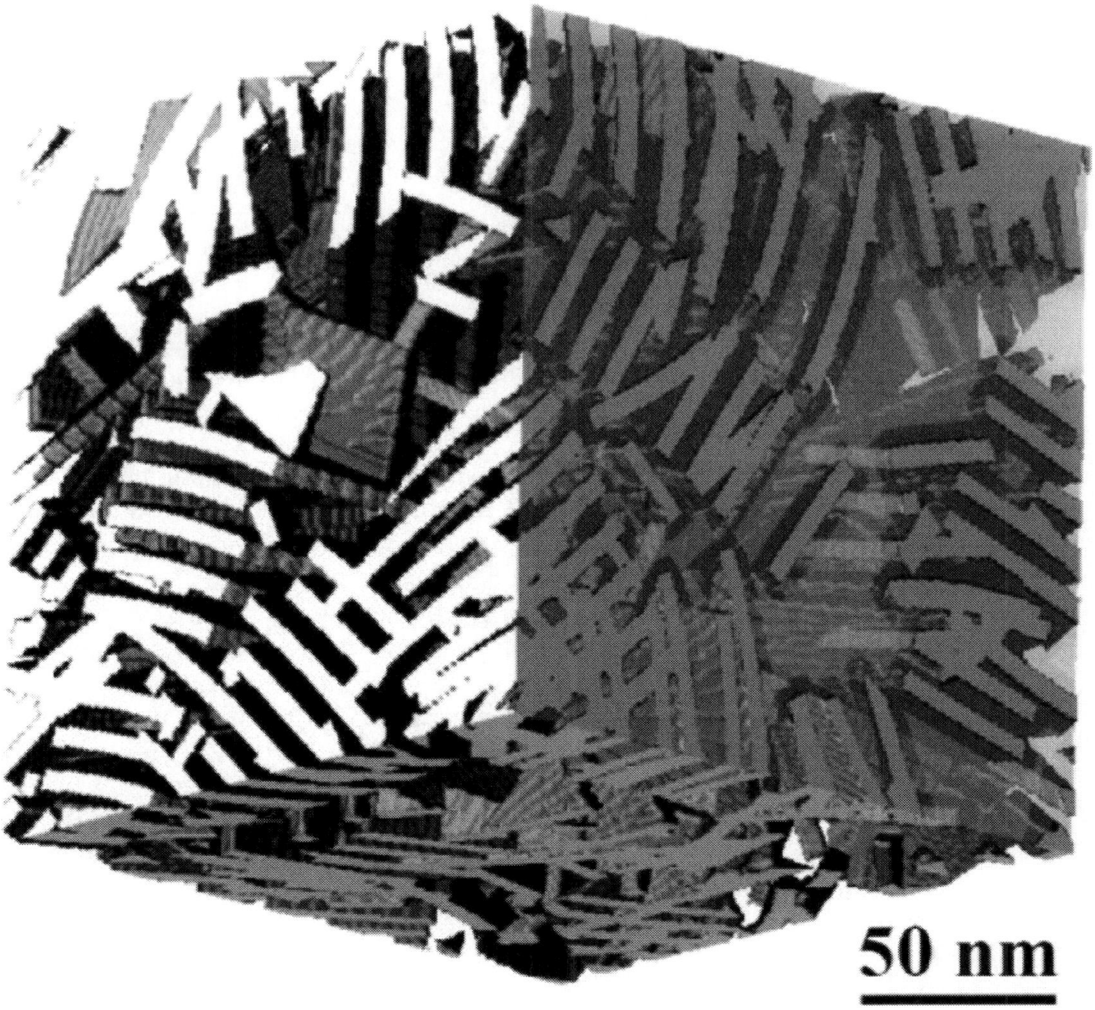


**Figure 1:** 3D image of semi-crystalline polyethylene. Grey parts represent crystalline phase, red is free amorphous phase and blue is constrained amorphous phase [4].

# 628223 Transport and Energy Processes in Polymeric Processing


**Abstract:**

Polymeric materials are used in many forms such as plastics, fibers, rubber, paints, coatings or adhesives and also as molded and fabricated articles implies that there must be a variety of ways in which the compounded resins can be processed and converted into finished products.

Polymers can be classified in different ways, as (i) Natural and synthetic polymers- natural polymers such as cotton, silk, wool and rubber,synthetic PE,PP,etc. (ii) Organic and inorganic polymers- a polymer whose backbone chain is made of carbon atoms is termed an organic polymer and some atoms attached to the side valancies of the backbone carbon atoms may be hydrogen, oxygen, nitrogen, etc. (iii) Thermoplastic and thermo-setting polymers- Polymers may soften on heating, reshaping, retaining the shape and stiffen on cooling and can be repeated several times are termed as thermoplastics e.g., PE, PP, etc. Some polymers, on the other hand undergo some chemical change on heating and convert themselves into an infusible mass, are called thermo-setting polymers e.g., yolk of egg. (iv) Elastomers- when polymers vulcanized into rubbery products exhibiting good strength and elongation, are termed as elastomers e.g., natural rubber, synthetic rubber, silicon rubber. (v) Fibers- polymers drawn into long filament- like materials, whose length is at least 100 times its diameter, are called fibers e.g., nylon and terylene. (vi) Resins- Polymers used as adhesives, coatings, etc., in a liquid form e.g., epoxy adhesives.

Molecular theories for polymeric liquids include the kinetic theories for polymers and can be divided roughly into two classes: network and single- molecule theories:

1. The network theories were originally developed for describing the mechanical properties of rubber. One imagines that the polymer molecules in the rubber are joined chemically during vulcanization. The theories have been extended to describe molten polymers and concentrated solutions by postulating an ever- changing network in which the junction points are temporary, formed by adjacent strands that move together for a while and then gradually pull apart. It is necessary in the theory to make some empirical statements about the rates of formation and rupturing of the junctions.
2. The single- molecule theories were originally designed for describing the polymer molecules in a very dilute solution, where polymer- polymer interactions are infrequent. The molecule is usually represented by means of some kind of "bead spring" model, a series of small spheres connected by linear or nonlinear springs in such a way as to represent the molecular architecture; the bead spring model is then allowed to move about in the solvent, with the beads experiencing a Stokes' law drag force by the solvent as well as being buffeted about by Brownian motion. Then from the kinetic theory one obtains the "distribution function" for the orientations of the molecules (modeled as bead spring structures); once this function is known, various macroscopic properties can be calculated.


**Polymeric melts**

The behavior of polymeric melts are

1. Non- Newtonian polymer melt
2. Linear and non- linear viscoelastic behavior
3. Isothermal and non- isothermal melt behavior

**Convective momentum transport**

Assumption is a volume element of the polymer melt $\Delta x\Delta y\Delta z$ of arbitrary shape.

The viscosity arises from a consideration of the random motion of the molecules in the fluid- i.e., random molecular motion with respect to the bulk motion of the fluid. Momentum can, in addition, be transported by the bul flow of the fluid, and this process is called *convective transport*.

Combined *momentum flux tensor* $\boldsymbol{\Phi}$, which is sum of the *convective momentum flux tensor* $\rho\mathbf{vv}$ and the *molecular momentum flux tensor* $\boldsymbol{\pi}$ and that the latter can be written as the sum of $p\boldsymbol{\delta} + \boldsymbol{\tau}$, $\mathbf{v}$ is the velocity vector of the fluid

$$\boldsymbol{\Phi} = \boldsymbol{\pi} + \rho\mathbf{vv} = p\boldsymbol{\delta} + \boldsymbol{\tau} + \rho\mathbf{vv}$$

**The equation of continuity**

To get the equation of continuity we write a mass balance over the volume element $\Delta x\Delta y\Delta z$, which describes the time rate of change of the fluid density $\rho$ at a fixed point in the fluid element. Polymer melt is assumed to be incompressible.

{ rate of increase of mass } = { rate of mass in } – { rate of mass out }

$$\frac{\partial \rho}{\partial t} = -(\nabla \cdot \rho\mathbf{v})$$

$\frac{\partial \rho}{\partial t}$ is the rate of increase of mass per unit volume

$-(\nabla \cdot \rho\mathbf{v})$ is the net rate of mass addition per unit volume by convection also called "divergence of $\rho\mathbf{v}$, $\rho\mathbf{v}$ is the mass flux.

For incompressible fluid = $(\nabla \cdot \mathbf{v}) = 0$

**The equation of motion**

To get the equation of motion we write a momentum balance over the volume element $\Delta x\Delta y\Delta z$,

{ rate of increase of momentum ] = { rate of momentum in } – { rate of momentum out } + { external force on the fluid }

$$\frac{\partial \rho\mathbf{v}}{\partial t} = -[\nabla \cdot \boldsymbol{\Phi}] + \rho\mathbf{g}$$

Where $\rho\mathbf{v}$ is the momentum per unit volume at a point in the fluid, $\rho\mathbf{g}$ is the external force per unit volume at a point in the fluid,

$\frac{\partial \rho \mathbf{v}}{\partial t}$ is the rate of increase of momentum per unit volume,

$-[\nabla \cdot \rho\mathbf{v}\mathbf{v}]$ is the net rate of momentum addition by convection per unit volume,

$-\nabla p - [\nabla \cdot \boldsymbol{\tau}]$ is the net rate of momentum addition by molecular transport per unit volume

$$\frac{\partial \rho \mathbf{v}}{\partial t} = -[\nabla \cdot \rho\mathbf{v}\mathbf{v}] - \nabla p - [\nabla \cdot \boldsymbol{\tau}] + \rho\mathbf{g}$$

**The equation of mechanical energy**

The equation of change of kinetic energy:

$$\frac{\partial}{\partial t}(1/2\rho v^2) = -(\nabla \cdot 1/2\rho v^2\mathbf{v}) - (\nabla \cdot p\mathbf{v}) - p(-\nabla \cdot \mathbf{v}) - (\nabla \cdot [\boldsymbol{\tau} \cdot \mathbf{v}]) - (-\boldsymbol{\tau} : \nabla\mathbf{v}) + \rho(\mathbf{v} \cdot \mathbf{g})$$

$-(\nabla \cdot p\mathbf{v})$ =net rate of work done by pressure of surroundings on the fluid,

$-p(-\nabla \cdot \mathbf{v})$= net rate of reversible conversion of kinetic energy into internal energy,

$-(\nabla \cdot [\boldsymbol{\tau} \cdot \mathbf{v}])$ = net rate of work done by viscous forces on the fluid,

$-(-\boldsymbol{\tau} : \nabla\mathbf{v})$ = net rate of irreversible conversion of kinetic energy into internal energy and reversible conversion to elastic energy.

$\rho(\mathbf{v} \cdot \mathbf{g})$ = net rate of work by external force on the fluid,

$-(\nabla \cdot 1/2\rho v^2\mathbf{v})$ = net rate of addition of kinetic energy by conversion per unit volume,

$\frac{\partial}{\partial t}(1/2\rho v^2)$= rate of increase of kinetic energy per unit volume

$\boldsymbol{\tau}$ =second-order stress tensor which is symmetric i.e.,$\tau_{ij} = \tau_{ji}$

$\boldsymbol{\Phi}$=potential energy per unit mass, defined by $\mathbf{g} = -\nabla\boldsymbol{\Phi}$

$$\frac{\partial}{\partial t}(1/2\rho v^2 + \rho\boldsymbol{\Phi}) = -(\nabla \cdot (1/2\rho v^2 + \rho\boldsymbol{\Phi}\,\mathbf{v}) - (\nabla \cdot p\mathbf{v}) - p(-\nabla \cdot \mathbf{v}) - (\nabla \cdot [\boldsymbol{\tau} \cdot \mathbf{v}]) - (-\boldsymbol{\tau} : \nabla\mathbf{v})$$

This is an equation of change for kinetic+potential energy and referred to as the equation of change for mechanical energy.

Solids, such as single noncubic crystals, fibrous materials, and laminates, are anisotropic. Heat flux vector **q** does not point in the same direction as the temperature gradient.

$\mathbf{q} = -[\boldsymbol{\kappa} \cdot \nabla T]$

in which **κ** is a symmetric second- order tensor called the thermal conductivity tensor.

**The energy equation**

The conservation of energy follows:

{rate of increase of kinetic and internal energy} = {net rate of kinetic and internal energy addition by convective transport} + {net rate of heat addition by molecular transport (conduction)} + {rate of work done on system by molecular mechanisms (i.e., by stresses)} + {rate of work done on system by external forces (e.g., by gravity)}

This may be written in vector notation as

$$\frac{\partial}{\partial t}(1/2\rho v^2 + \rho U) = -(\nabla \cdot \mathbf{e}) + \rho(\mathbf{v} \cdot \mathbf{g})$$

The combined energy flux vector **e** as follows:

$$\mathbf{e} = (1/2\rho v^2 + \rho U)\,\mathbf{v} + [\boldsymbol{\pi} \cdot \mathbf{v}] + \mathbf{q}$$

where

$(1/2\rho v^2 + \rho U)\,\mathbf{v}$ =convective energy flux vector

$[\boldsymbol{\pi} \cdot \mathbf{v}]$=molecular work flux vector

**q**=molecular heat flux vector

**e**=combined energy flux vector

The equation of energy,

$$\frac{\partial}{\partial t}(1/2\rho v^2 + \rho U) = -(\nabla \cdot (1/2\rho v^2 + \rho U)\,\mathbf{v}) - (\nabla \cdot \mathbf{q}) - (\nabla \cdot p\mathbf{v}) - (\nabla \cdot [\boldsymbol{\tau} \cdot \mathbf{v}]) + \rho(\mathbf{v} \cdot \mathbf{g})$$

Where

$\frac{\partial}{\partial t}(1/2\rho v^2 + \rho U)$ =rate of increase of energy per unit volume

$(\nabla \cdot (1/2\rho v^2 + \rho U)\,\mathbf{v})$ =rate of energy addition per unit volume by convective transport

$(\nabla \cdot \mathbf{q})$ =rate of energy addition per unit volume by heat conduction

$(\nabla \cdot p\mathbf{v})$ =rate of work done on fluid per unit volume by pressure forces

$(\nabla \cdot [\boldsymbol{\tau} \cdot \mathbf{v}])$ =rate of work done on fluid per unit volume by viscoelastic forces

$\rho(\mathbf{v} \cdot \mathbf{g})$ =rate of work done on fluid per unit volume by external forces

# 624759 Modeling and Simulation of Fixed-Bed Reactors Made of Open-Cell Metallic Foam Pellets

**Abstract:**

## Introduction

Fixed-bed catalytic reactions – exothermic or endothermic – are widely used in the chemical and process industries. They are effectuated by the slender tubes filled with catalyst pellets and enwrapped with the heat transfer media, e.g. circulating gas, to either supply or remove the heat of reaction.[1] To guarantee stable operation and high productivity in a commercial-scale setup, the selection of an adequate catalyst pellet type is of prime important, which is based on the needs such as lower pressure drop, improved heat transfer and high mechanical stability. Indeed, these requirements have been identified as of contrasting nature, therefore, challenging to satisfy the entire demands with the commonly used ceramic type of pellets.[2]

In recent years, the advancement in the manufacturing technology has triggered the production of alloyed metallic foam pellets (see Fig. 1a), which are stable even in the high temperature environment.[3] Due to the superior properties as high porosity, good mechanical stability, global thermal conductivity and the like; the metallic foam pellets have been considered as a beneficial alternative to the conventional ceramic pellets, since they encourage less energy consumption and process efficiency, by virtue of lower pressure drop and improved heat transfer. The methods of process enhancement in the fixed-bed reactor should be encouraged, as it is heavily employed in many relevant large-scale processes, e.g. production of hydrogen by the SMR process. Hydrogen as a green energy is being prefigured as the solution to many issues related to climate change.

However, the tortuous inner structure of metallic foam pellets added with considerable internal flow causes a complex flow field in a packed bed arrangement compared to non-porous pellets. Additionally, for the case of slender packed beds, the reactor wall imparts particle-ordering-effect in the form of variable radial void fraction with the highest bed voidage very close to the wall, followed by a pattern of several maxima and minima from the reactor wall to the center core of the bed. The transport phenomena as well as heat transfer mechanism are dependent on the local flow behavior, which is regulated by the catalyst shape and its arrangement in a packed bed structure. Therefore, it is important to optimize the shape of the metallic foam pellet that is suitable for particular reaction conditions. Despite the fact, a systematic study pertaining to fixed-beds made of metallic foam pellets is still missing, which discourage the manufacturers for its mass production. Recently, Wehinger et al.[4] proposed a novel particle-resolved CFD approach to analyze the fixed-bed reactors composed of metal foam pellets. The objective of this contribution is to develop a detailed CFD workflow to investigate the flow characteristics and the heat transfer performance of a packed bed made of metal foam pellets. This modeling framework can act as a suitable plug-in of a Digital Twin, which provides a virtual environment to explore the adequacy of any pellet shapes before any decisions concerning manufacturing; reduces the risk in costly investments.

## Methods

The particle-resolved CFD approach that accounts for the actual bed geometry was used. The fluid flow throughout the interstitial regions was fully resolved, whereas flow through the porous foam pellets was considered by the porous-media model. The pressure drop at each pellet level was modeled as per the Lacroix equation[5], and the effective thermal conductivity by the Schuetz-Glicksman model.[6] A realistic bed structure was generated with the aid of Rigid Body Dynamics algorithm, integrated with the animation software Blender. The CFD simulations were carried out with Simcenter STAR-CCM+ from Siemens Industry Software Inc. Fig. 1c illustrates the proposed CFD workflow.

A suitable experimental setup was also realized to validate the CFD model, as schematically shown in Fig.1b. The steel reactor has an inner diameter of 69 mm and a height of 2.5 m. The pre-heated Nitrogen gas was fed to the top of the reactor, and is cooled while flows through the bed by the cooling oil circulating along the reactor wall. A thermowell equipped with 16 thermocouples was placed at the center core of the bed to measure the axial temperature from the top of the bed to the bottom. The experiment was carried out at different combinations of flow rates and inlet gas temperature, $250 < Re_p < 2500$. The temperature data corresponding to the steady-state condition were recorded.

## Results and Discussion

Figs. 2a and 2c show the local velocity field and the temperature distribution in a vertical sectional plane, respectively. The local rise in the velocity along the interstitial regions, and the channelling effects near the wall – reactor and thermowell – are clearly visible. As the porous foam pellets allow flow through the pellets, the local velocity fluctuations in the foam packed beds are lower compared to the solid pellets (not shown here). The temperature distribution in a packed bed is dependent on a number of factors such as effective pellet conductivity, lateral convective mixing, magnitude of flow channelling near the wall, etc. Moreover, the significance of each heat transfer mechanisms – conduction, convection and radiation – depends on the volumetric flow rate and the operating temperature. Fig. 2d shows the comparison of the measured and the predicted axial temperature along the center-axis of the bed, at a flow rate of 45 $nm^3/h$ and the inlet gas temperature of about 378 K. The simulated temperature profile shows an excellent agreement with the experimental data. As shown in Fig. 2c, the predicted pressure drop is also in good agreement with the experiment. To come up with an absolute CFD model, the validation process will be carried out for a wide range of operating conditions, and appropriate tuning in the modeling approach will be made, if necessary. Subsequently, the validated CFD model will be used to explore the optimal shape of metal foam pellet.

## Conclusion

An adapted particle-resolved CFD approach integrated with Rigid Body Dynamics was used to virtually explore the transport phenomena in a packed-bed made of metallic foam pellets. The simulated results are validated against the axial temperature profile and the pressure drop measured in an experimental reactor setup. The proposed CFD model has found as adequate to investigate the flow characteristics and the heat transfer performance of the foam beds. The ability to obtain accurate predictive results in a manageable timeframe is promising, which allows the exploration of the optimal metal foam pellet shape by achieving a reasonable trade-off between pressure drop and heat transfer performance.

## References


1. Jurtz N, Flaischlen S, Scherf SC, Kraume M, Wehinger GD. Enhancing the Thermal Performance of Slender Packed Beds through Internal Heat Fins. Processes. 2020; 8(12):1528.
2. Dixon AG. CFD study of effect of inclination angle on transport and reaction in hollow cylinder catalysts. Chemical Engineering Research and Design. 2014;92(7):1279-1295.
3. Walther G, Klöden B, Büttner T, et al. A New Class of High Temperature and Corrosion Resistant Nickel-Based Open-Cell Foams. Adv. Eng. Mater. 2008;10(9):803-811.
4. Wehinger GD, Kolaczkowski ST, Schmalhorst L, Beton D, Torkuhl L. Modeling fixed-bed reactors from metal-foam pellets with detailed CFD. Chemical Engineering Journal. 2019;373:709-719.
5. Lacroix M, Nguyen P, Schweich D, Pham Huu C, Savin-Poncet S, Edouard D. Pressure drop measurements and modeling on SiC foams. Chemical Engineering Science. 2007;62(12):3259-3267.
6. Ranut P. On the effective thermal conductivity of aluminum metal foams: reviewand improvement of the available empirical and analytical models. Appl. Therm. Eng. 101 (2016) 496–524.

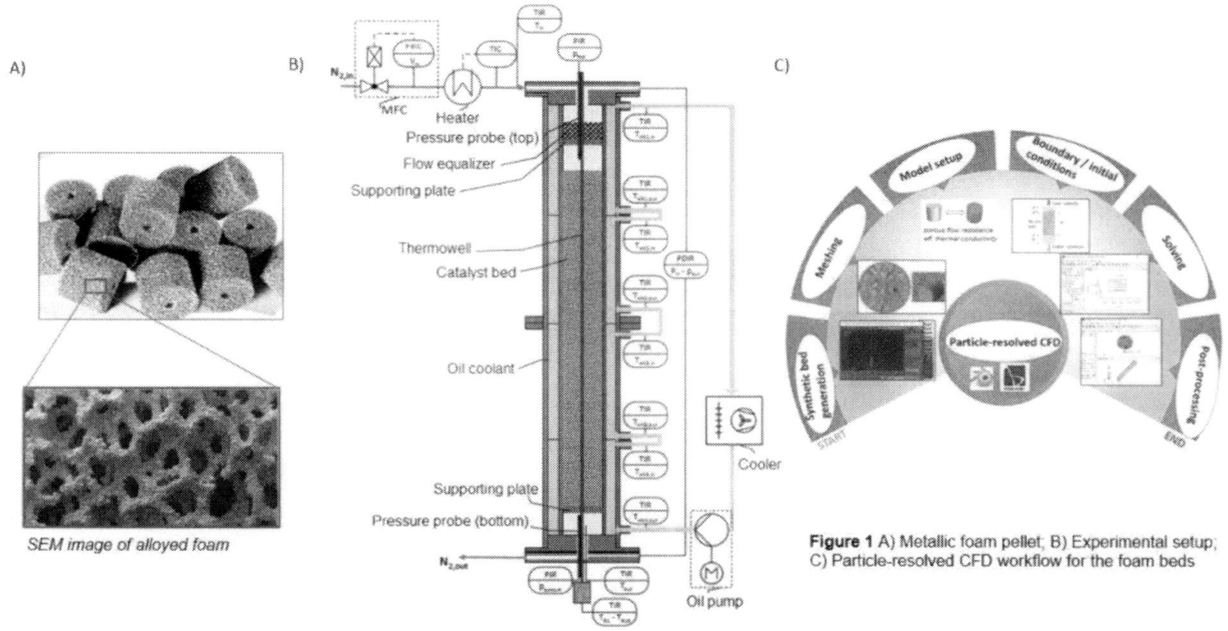


**Figure 1** A) Metallic foam pellet; B) Experimental setup; C) Particle-resolved CFD workflow for the foam beds

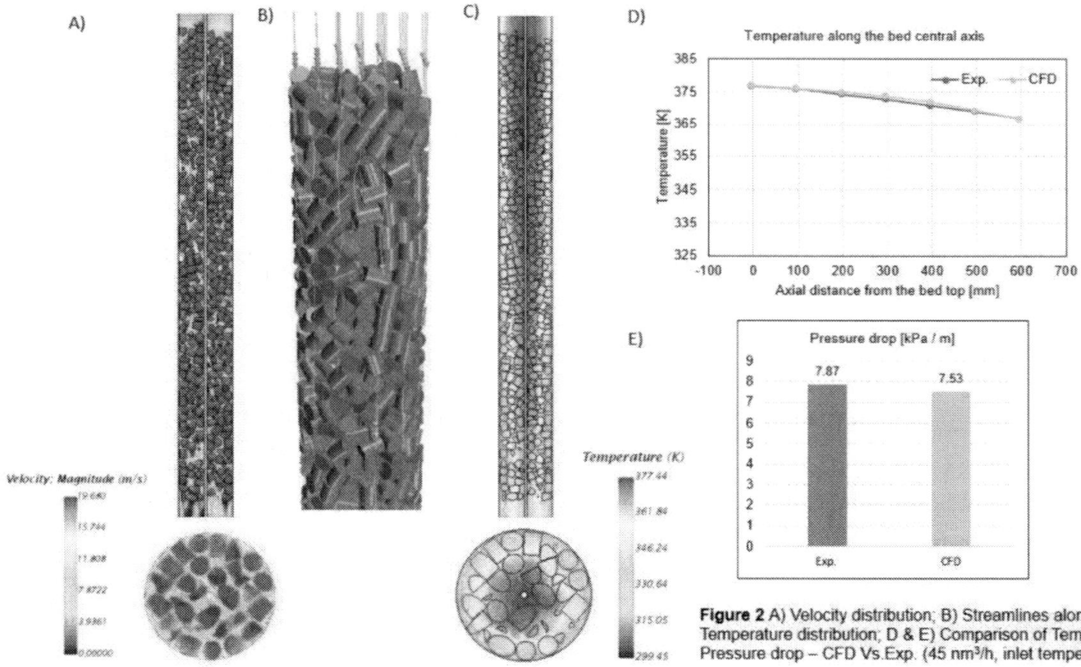


**Figure 2** A) Velocity distribution; B) Streamlines along the bed; C) Temperature distribution; D & E) Comparison of Temperature profile and Pressure drop – CFD Vs.Exp. (45 $nm^3/h$, inlet temperature = 378 K)

# 624091 Catalyst Dilution Criteria for Mass Transfer Intensification in Lab Gas-Liquid Fixed-Bed Reactors

**Abstract:**

Fixed-bed reactor technology at the lab scale has experienced a strong evolution during the past half century. Lab reactor volumes have been substantially reduced for sustainability reasons. This significant reduction has been accompanied by the development of new generations of catalytic formulations with improved activity, selectivity and stability. Reactor downscaling coupled with increasing chemical reaction rates have triggered many studies on fixed-bed hydrodynamics and mass transfer phenomena. As a matter of fact, these phenomena can become limiting at lab scale, and as a consequence, they can prevent from a reliable catalyst performances scale-up.

Some approaches for reducing fixed-bed reactor effects on kinetics measurements can be distinguished. Bed dilution with inert particles is an efficient way to improve reactor hydrodynamics and heat transfer. Several studies in literature show that adding an inert material to fill the catalytic bed porosity strongly reduces axial dispersion effects and prevents from preferential pathways[1-3]. This technique intensifies heat transfer providing that the inert material thermal conductivity is sufficiently high, as with SiC or non-porous ceramics (alumina, zirconia, etc)[4]. Literature on catalyst dilution for the improvement of mass transfer within fixed-bed reactors at the lab scale is poor. Different ways of diluting catalyst can be considered. The inert material can be uniformly distributed along the entire bed. Different dilution factors can be applied at different reactor locations depending on the expected kinetic rate evolution. Another approach for catalyst dilution is to alternate catalytic and inert beds. What is the best technique to be applied for a specific case? Where should the inert material be placed? In what proportion?

The aim of this work is to study the effect of different catalyst dilution techniques on kinetics measurements related to chemical systems limited by gas-liquid mass transfer. More particularly, this study must provide catalysts suppliers and experts from the field of heterogeneous catalysts kinetic modeling with specific guidelines for improving catalyst testing methods.

The methodology, based on global phenomena and dilution modeling to assess and optimize reactor loading techniques, was applied to the oil residue hydrodemetallation and to the benzene hydrogenation. A kinetic parameters regression with available experimental data was carried out for both the applications prior to the evaluation of the different dilution approaches. Splitting consecutive catalytic beds with an inert bed of SiC particles with the same size as the catalyst pellets poorly improves gas-liquid mass transfer. Uniform dilution consisting on directly mixing both the catalyst and the inert material allows to strongly intensify gas-liquid mass transfer. This is due to the reduction of the reactant consumption per reactor unit volume, whereas the gas-liquid mass transfer rate is kept constant. Combinations of both abovementioned techniques can be useful for high reaction rate systems operated at conversions higher than 70%. The results obtained in this work allowed the establishment of a criterion on the minimum dilution factor to be applied for neglecting gas-liquid mass transfer effects on kinetics measurements. This criterion depends on conversion degree, external mass transfer rate as well as on global pseudo-second order kinetics. The dilution factor, defined as the ratio between the volume of catalyst and the total volume occupied by both the catalyst and the inert particles, decreases with the decrease of conversion and with the increase of the pseudo-second order kinetic constant. The agreement between rigorous modeling and criterion results is excellent, as illustrated in the figure below.

# References

[1] Sie S.T. (1996) Miniaturization of hydroprocessing catalyst testing systems: Theory and practice, *AIChE J.* **42**, 12, 3498-3507.

[2] Mears D.E. (1971) The role of axial dispersion in trickle-flow laboratory reactors, *Chem. Eng. Sc.* ***26**, 1361-1366.*

[3] Gierman H. (1988) Design of laboratory hydrotreating reactors: Scaling Down of Trickle-flow Reactors, *Appl. Catal.* **43**, 277-286.

[4] Taniewski M., Lachowicz A., Skutil K., Czechowicz D. (1996) The effect of dilution of the catalyst bed on its heat-transfer characteristics in oxidative coupling of methane, *Chem. Eng. Sc.* **51**, 18, 4271-4278.

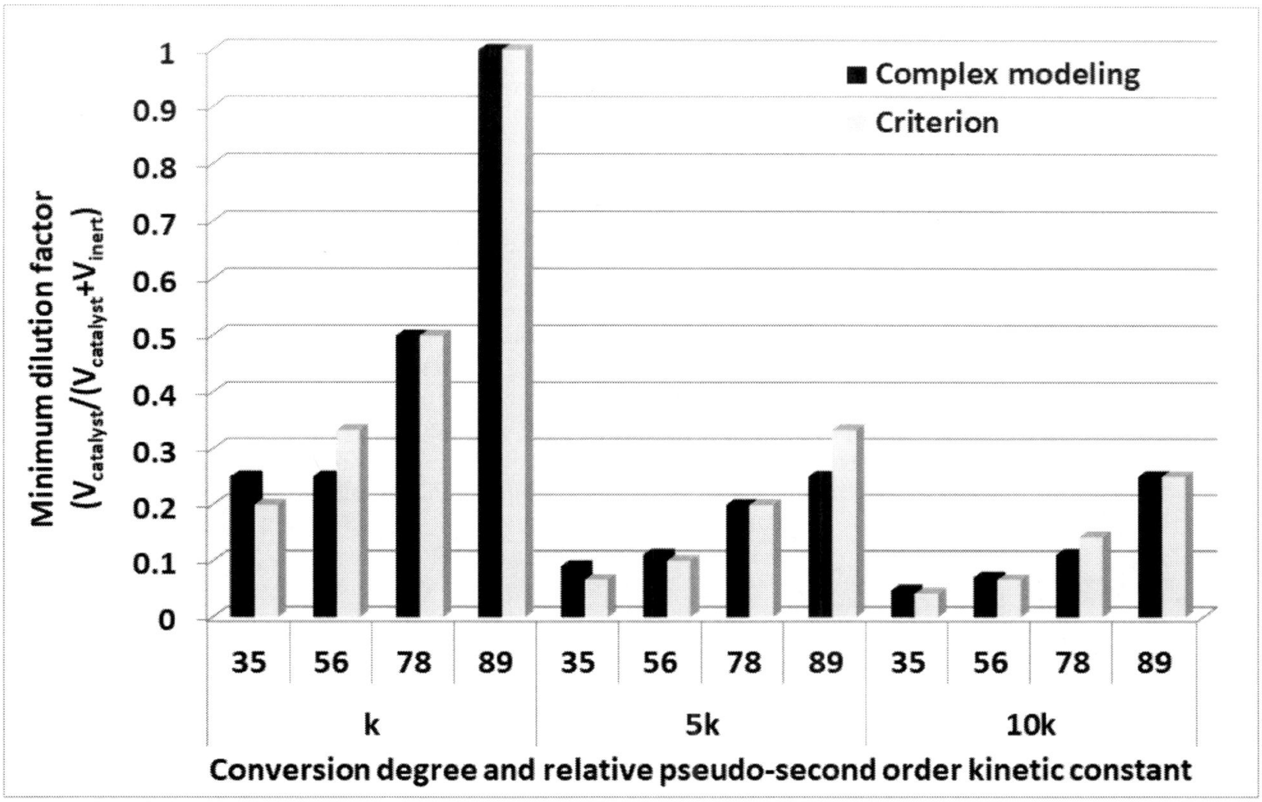

# 630939 Evaluation of Real Time, Inline Characterization of Bubble Size and Shape Distributions in Highly Concentrated Dispersions

**Abstract:**

Today many different techniques for sizing transient fluid particle behavior in production vessels are available. Often samples are withdrawn over time, which are later diluted or stabilized, prior to their measurements. These sampling techniques neither guarantee that the drop or bubble sizes are frozen, nor that they are preserved during the sampling and have been criticized by many authors. Especially for technical applications with fast reactions and strong coalescence such kind of measurements are not suitable. Even for sampling times less than one second, the drastic change in the flow condition during sampling results in significant measurement errors. The control of the fluid particle size in such systems is an important goal for method development.

methods

Image-based set-ups are capable of a simultaneous analysis of the size and the shape of the bubbles and they are the number one technology in the industrial as well the academic world. Static and dynamic image analysis have been evaluated by the ISO group to standardize the multiple procedures. The results are given in the ISO13322-1 and ISO13322-2. Image analysis is a technique that has gained popularity in different applications. Two different set-ups have been studied to evaluate the capability of inline imaging connected with an automated image analysis software: a mixing vessel as well as a bubble column. Different image analysis algorithms and their results based on the named experiments are evaluated and categorized based on their advantages as well as disadvantages. These algorithms range from classical computer vision approaches to machine learning and artificial intelligence applications.

**results**

The influence of the stirrer geometry, the stirrer speed, the air flow rate as well as the air pressure on the bubble size distribution have been studied extensively. The different experiments do give a broad overview on possible process conditions in terms of Reynolds and Weber number. The different algorithms used give a greater understanding of the development and challenges in the application of image analysis in those circumstances. The following figure shows a typical image analysis result using a pattern matching algorithm. Despite the high hit rate and acceptable degree of precision, the image already shows the limitation of this algorithm if the shape of the object of interest varies from circular objects. For thes more challenging analysis task, as the degree of freedom is increasing, machine learning approaches have been applied with astonishing results. The presentation will show these results and the necessary preparations to achieve real time feedback from such a highly concentrated dispersion.

**Fig. 1 - Example images of the detection results. The figures show the original image with the detection results (upper image) and the filtered and sharpend image (lower images). The results were obtained for the GDX-impeller at a stirrer speed of 1300 rpm.**

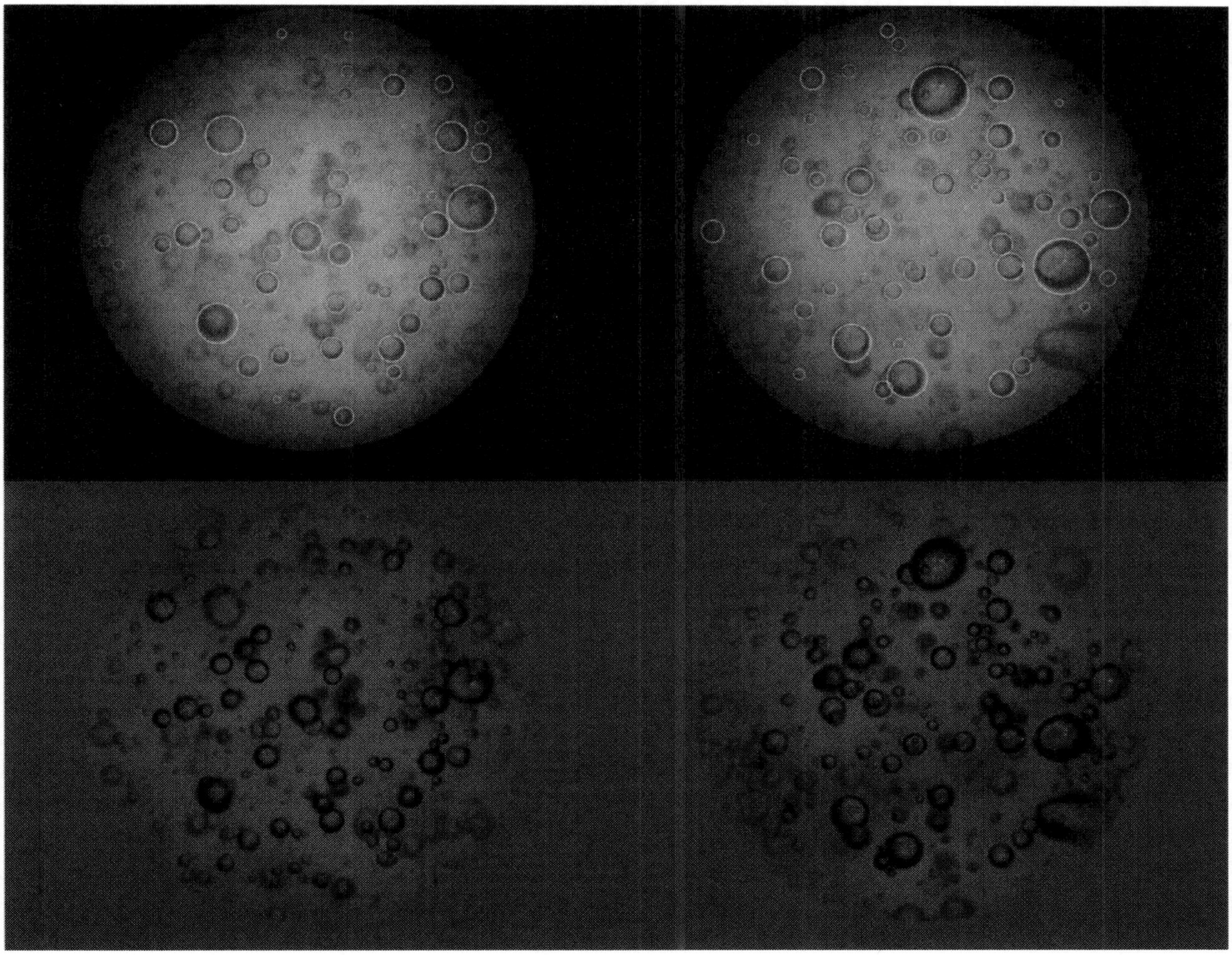

# 623413 Experimental Investigation of the Impact of Tubes Arrangement on Local Heat Transfer Coefficient in Fluidized Bed Reactor for Fischer-Tropsch Synthesis

**Abstract:**

The process of converting biomass, natural gas, and coal to liquid fuels and chemicals via Fisher Tropsch synthesis is currently a major interest of the energy and chemical industries, representing a valuable addition to expanding fuel and product resources. The fluidized bed reactor is the preferred reactor for carrying out the Fischer Tropsch process's exothermic reaction. Excess heat due to exothermicity of reaction is removed by inserting a bundle of cooling tubes into this reactor, thus controlling the reaction temperature. Designing and scaling-up this reactor is still a problematic engineering task due to the absence of reliable mathematical models to predict heat transfer for this reactor when equipped with a bundle of heat exchange tubes. Developing and improving any mathematical model or simulation requires reliable data, which is lacking for this kind of reactor. Therefore, this study aims for the first time to assess and quantify the influence of the presence of vertical tubes and their arrangement on the local heat transfer coefficient by using an advanced heat transfer technique. To achieve this aim, a plexiglass column with a diameter of 0.13 m and a height of 1.83 m was designed and fabricated. A bundle of thirty stainless steel tubes was inserted vertically into the fluidized bed column. This bundle of vertical tubes was covered 25% of the cross-sectional area of the column to simulate the industrial Fischer-Tropsch reactor. Two types of tube arrangements (i.e., square and triangle pitch) were investigated in this study. Additionally, two different particles, namely glass beads and aluminum oxide, were used in this investigation. The specifications of those solid particles were chosen according to Fischer-Tropsch industrial requirements. Measurements of the local heat transfer coefficient were measured at different radial and axial locations under a wide range of superficial gas velocities (0.02 to 0.4 m/s). Furthermore, a two-dimensional (2D) image was created by measuring the local heat transfer coefficient at thirty locations at an axial level of 13 cm from the gas distributor. This 2D image visualizes local heat distribution over the entire cross-sectional column.

The knowledge and findings gained from this study would further improve the fundamental understanding of the impact of vertical tubes on heat transfer in fluidized bed reactors. Additionally, the data will also be used to verify reactor models and CFD codes and simulations that promote such a reactor's design and scale-up purposes. The results of this work will be presented in detail on the conference day.

# 629887 Thermochemical Storage of Solar Energy Via Calcium Looping: Kinetic Study and Reactor Modeling of a Solar Calciner

**Abstract:**

## Introduction

Heat from solar radiation is currently collected in concentrated solar power (CSP) plants where it is converted to electricity via a power cycle. The intermittent nature of solar energy necessitates the development of efficient storage systems for the on demand power supply. Commercially, the heat is converted to electricity using molten salts as the working and storage medium, which suffer from several drawbacks such as temperature limits, low efficiency and short time storage [1]. Calcium looping (CaL), which relies on the cyclic reaction of CaO with $CO_2$, is potentially a means for storing solar energy in a chemical substance by driving the endothermic $CaCO_3$ decomposition. The reverse carbonation reaction can then produce heat ($\Delta H^o$=-178 kJ/mol) at high temperatures (~850 °C under pure $CO_2$ atmosphere) increasing the power cycle's efficiency, while the regeneration of $CaCO_3$ is feasible within the operational temperature range of CSP towers [2–4]. These features, along with the abundance of materials, high energy density and lack of toxicity render CaL a promising route for storing solar energy in the long term. $CaCO_3$ decomposition reaction has been studied for many years, but consensus has not been achieved regarding the rate expression and the effect of $CO_2$ partial pressure [5]. The existence of a robust kinetic model is of high importance for the reactor design. Additionally, the design of a calciner that efficiently converts solar to chemical energy through $CaCO_3$ decomposition is probably the greatest challenge of this concept. Herein, a lab-scale experimental kinetic study of limestone decomposition was carried out targeting to the development of a reliable kinetic model that will explain the literature discrepancies. This kinetic model was then used for the design of a 1-D drop-down reactor that can be used for thermochemical storage of solar energy in CSP plants.

## Methodology

Kinetic experiments of limestone (Granicarb 0.1/0.8, OMYA S.A.) decomposition were conducted in a continuous flow fixed-bed reactor at temperatures between 825-885°C and $CO_2$ partial pressures <0.6 atm. Parameters such as the particle size (45-75 μm), the gas hourly space velocity (22500 $h^{-1}$) and the bed weight and concentration (0.1 g of sample diluted in 1.5 g of quartz) were defined experimentally such as to minimize the effect of heat and mass transfer phenomena on the observed rate. For each pair of temperature and $CO_2$ partial pressure conditions, an experimental conversion - time curve was derived.

Calcination is a reaction that takes place at the surface of the reactant $CaCO_3$ which is reduced during reaction. Accordingly, the grain model was used that predicts the conversion evolution assuming that the particle is consisted of impervious spherical grains that are shrinking during the course of reaction (Figure 1A). The grain model equation was fitted in each of the experimental conversion-time curve deriving the specific reaction rate $r\ (T,\ P_{CO2})$ (e.g the rate of reaction per unit of reactant surface area in $m^3/m^2s$ or m/s) as the only fitting parameter. The expression of $r$ which is a function only of reaction temperature and $CO_2$ partial pressure is discussed in the results section.

The type of the calciner modeled herein was a 1-D plug-flow drop-down narrow tube reactor. Limestone particles and the gas mixture ($CO_2/N_2$) enter from the top of the reactor and decomposition takes place as they move downwards. The reactor was discretized in the axial direction (1000 slices) assuming homogeneous $P_{CO2}$ and temperature conditions inside each slice (Figure 1B). The residence time of the solids in each slice was calculated assuming Stokes flow around the particles (Re < 1) [6]. Additionally, the limestone particles were discretized in the radial direction (100 shells) to take into account the intraparticle $CO_2$ partial pressure gradients (Figure 1A). Thus, $P_{CO2}$ was a function of radial position inside the particles and its expression was found by solving a differential equation assuming steady state conditions. The calculated solids residence time and the $P_{CO2}(r)$ were then introduced in the kinetic model for the estimation of the conversion change $\Delta X$. The released $CO_2$ and the heat absorbed due to the endothermicity of $CaCO_3$ decomposition were taken into account in the mass and energy balances of the slice, providing the input information of the subsequent slice volume. The same calculations were carried out for the next volume slices. Both the kinetic and reactor modeling were performed using the Matlab software (Mathworks).

## Results

From the kinetic experiments, Figure 1D shows the experimental values (points) and the grain model fitting (line) of conversion versus time for decomposition of limestone at 885°C and under different $CO_2$ partial pressures. The model fits adequately on the experimental data deriving one value of the only fitting parameter $r$ (rate per surface area unit) for each $X$-$t$ (the $r$ values are presented in Figure 1E). Also, the reaction rate is decreased when increasing $P_{CO2}$ as expected due to approach to equilibrium. The effect of $P_{CO2}$ on the specific reaction rate is non-linear and can be described well by a two stage mechanism (Figure 1C). According to this mechanism, $CO_2$ is firstly in an "adsorbed-like" stage before released in the gas phase in a second step. Structural changes in the solid phase may also happen simultaneously as proposed in the literature regarding the presence of a metastable CaO* phase as intermediate [7]. Assuming that the first step is controlling the rate ($E_a$=217 kJ/mol) and the second is at equilibrium, a Langmuir-type relation is derived which describes well the experimental values of the rate (lines in Figure 1E). Regarding the literature differences, it was concluded that they are mainly derived by heat and mass transfer effects, as well as, sintering phenomena that reduce the initial value of limestone's surface area leading to underestimation of rate values.

As described in the methodology section, the reactor model consists of three different sub-models, namely the reaction kinetic model, the hydrodynamic model and the intraparticle diffusion model. There are not many experimental data in the literature that meet the conditions for verifying the validity of all the three sub-models simultaneously. Fernandez et al. [8] carried out experiments in a pilot scale drop down tube reactor for calcining fine limestone particles. They imposed high gas/solids flow ratio to maintain constant $P_{CO2}$ and temperature conditions throughout the whole reactor. Their results were used for validating the reactor model with respect to the reaction kinetics but also the hydrodynamic sub-models, as the reaction time is related to the residence time of the solids inside the reactor. The intraparticle diffusion model could not be tested under such conditions as intraparticle $P_{CO2}$ gradients are minimized for very small particles. Figure 1F shows that the model predictions are in good agreement with the experimental data with some deviations probably related to the model approximations but also to experimental errors. Experiments carried out in our fixed-bed apparatus using different particle sizes for defining the size that minimizes the intraparticle diffusion resistance for the kinetic experiments are shown in Figure 1G. These experiments were used to validate the intraparticle diffusion along with the reaction kinetic sub-models (lines in Figure 1G). The model predicts that particles with diameter larger than 100 μm will exhibit slower reaction rates due to intraparticle $CO_2$ pressure gradients, which is in line with the experimental results. Again, the minor discrepancies observed may have been produced by the model assumptions (such as the adopted tortuosity value etc.). Apart from simulations for validation purposes, the model predicted that calcination of limestone can be achieved in residence times of a few seconds under relevant conditions for thermochemical energy storage.

## Conclusions

A kinetic study of limestone decomposition was carried out experimentally and theoretically. It was found to follow a two-stage mechanism implying a non-linear effect of the $CO_2$ partial pressure on the reaction rate. The developed kinetic model was subsequently used for the design of a 1-D drop-down narrow tube calciner that can be used in thermochemical energy storage applications in CSP plants. The validity of the kinetic and reactor models was verified after deriving satisfactory predictions on experimental data. The existence of robust models is strongly believed to enhance the maturity level of the application of calcium looping for the efficient exploitation of solar energy.

**References**

[1] S.E.B. Edwards et al., Sol. Energy, 86 (2012) 2494–2503.

[2] C. Ortiz et al., ACS Sustain. Chem. Eng., 6 (2018) 6404–6417.

[3] A. Scaltsoyiannes et al., Chem. Eng. J., (2020) 127207.

[4] E. Karasavvas et al., Energies, 13 (2020) 5905.

[5] A. Scaltsoyiannes and A. Lemonidou, Chem. Eng. Sci. X, (2020) 100071.

[6] P. Lisbona et al., Renew. Energy, 156 (2020) 1019–1027.

[7] J.M. Valverde et al., Phys. Chem. Chem. Phys., 17 (2015) 21912–21926.

[8] J.R. Fernandez et al., React. Chem. Eng., 4 (2019) 2129–2140.

**Acknowledgments**

This work has received funding from the European Union's Horizon 2020 research and innovation program under grant agreement No 727348, project SOCRATCES.

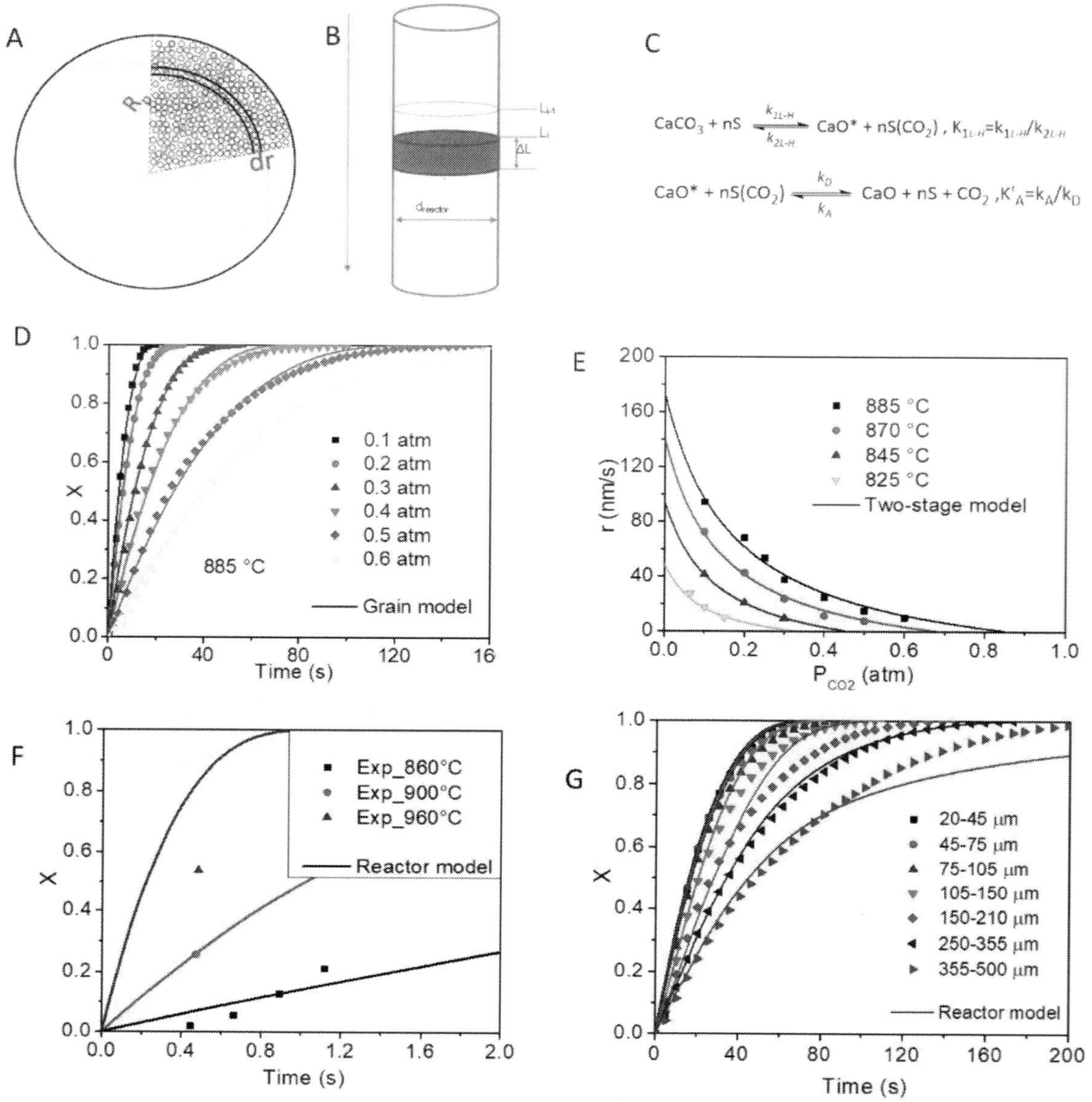


Figure 1. A) Representation of the grains that each limestone particle is consisted of according to the grain model and the radial discretization, B) axial discretization of the drop-tube reactor, C) the two-stage mechanism that describes the $CaCO_3$ decomposition reaction, D) experimental conversion-time data fitted by the grain model expression (solid lines) for calcination reaction at 885°C and under various $CO_2$ partial pressures, E) The values of specific rate of reaction *r* as a function of temperature and $P_{CO2}$ fitted by the rate expression derived by the two stage mechanism of Figure 1C, F) validation of the developed model on the experiments in a pilot scale reactor by Fernandez et al. [8], G) validation of the developed model on the experiments in a fixed-bed reactor using various particle sizes.

# 629695 Molecular Origins of Phase Stability in Model Phase-Change Nano-Emulsions for Thermal Energy Storage Elucidated By NMR Spectroscopy

**Abstract:**

Phase Change Materials (PCMs) are latent heat storage materials that can store or release large quantities of energy while undergoing thermodynamic phase transitions. Organic PCMs are of particular interest of large-scale applications because they are low cost, melt congruently, and exhibit good nucleating properties. Their major disadvantage is low thermal conductivity. By forming PCM nano-emulsions in which the organic PCMs are emulsified in water in the presence of surfactant, thermal conductivity increases significantly as well as the transport property does. However, PCMs nano-emulsions often become unstable due to repeated melting and freezing cycles, which can further be affected by shear flow during heat transfer applications. The surfactant distribution and dynamics upon thermal cycling and shear are poorly understood, but correlate with observed phase and heat transfer instabilities during their application.

To better understand the molecular behavior of surfactants upon thermal cycling and how it affects the stability of PCMs nano-emulsions, we designed a molecularly simple PCM nano-emulsion as a model system using octadecane as an oil, stearic acid as a surfactant and pH-adjusted water as the continuous medium. Solution-state $^{1}$H and $^{13}$C nuclear magnetic resonance (NMR) experiments were applied to a PCM nano-emulsion synthesized using stearic acid, half of which was uniformly $^{13}$C-labeled. The liquid content of octadecane within the nano-emulsions was measured quantitatively during melting and freezing, revealing a wide temperature range of solid-liquid co-existence that appears to be linked to phase instability. Interestingly, quantitative single-pulse $^{13}$C NMR experiments establish that the carbonyl surfactant head groups are in two different electromagnetic environments. $^{13}$C longitudinal ($T_1$) measurements also reveal two different $T_1$ relaxation times for the surfactant head groups. After repeated thermal cycling, $^{1}$H signal intensities of surfactant head groups decrease (e.g., by 33% over 120 cycles), indicating that the surfactant head groups lose molecular mobility, a key finding whose origins are still under investigation. $^{1}$H pulsed-filed-gradient (PFG) NMR methods were performed to measure the diffusion coefficients of the octadcane PCM nano-emulsions upon thermal cycling. $^{1}$H chemical shift, relaxation, and diffusion data were measured over different temperatures and number of thermal cycles, which are analyzed as potential indicators of phase stability. To generalize our findings, we synthesized a more complicated PCM nano-emulsion system by replacing stearic acid with Tween and Span mixed surfactants. The above NMR experiments were performed and the results were compared with the stearic acid model system. Overall, the results lay the groundwork towards molecular-level understanding of phase stability under thermal cycling in oil-water-surfactant PCM nano-emulsions for thermal energy storage.

# 623246 Pilot-Plant Testing of a Novel Phase-Change Solvent for $CO_2$ Capture

**Abstract:**

Adverse consequences of climate change are already evident and could negatively affect human health [1], living status [2] and global economy [3]. The development of $CO_2$ capture and storage facilities is considered as a key contributor to achieve the goals of the Paris Agreement [4], and is characterized as an essential tool to decarbonize the heavy industry [5]. Post-Combustion Capture (PCC) with reactive absorption is a mature option that is suitable for various different industrial sectors such as coal and natural gas power plants, cement plants and so forth [6]. However, the high energetic requirements of 4.0 GJ/ton $CO_2$ incurred by conventional solvents such as MEA (monoethanolamimne), prohibit the wide deployment of this technology at industrial scale.

A new type of solvents called phase-change solvents (PCSs), promise up to 50% reduction of capture costs with experimentally recorded regeneration energies of 2.3-2.5 GJ/ ton $CO_2$ [7]. This is achieved through a liquid-liquid phase split, triggered by reaction with $CO_2$ and/or temperature increase [8], that replaces part of the energy intensive treatment in the desorber by mechanical separation of the liquid phases and direct recycle of one of them to the absorber. While several PCSs have been proposed in published literature and undergone lab-scale characterization [7], there are very few PCSs that have been tested at pilot plant scale (i.e., DMCA/MCA/AMP [8], DEEA/MAPA [9], DMX [10] and 3H [11]). Pilot plant testing is a critical step in the course of process upscaling as the validity of the lab-scale results is tested and issues that may not appear at lab-scale are uncovered. Recently, Papadopoulos et al. [12] proposed the novel PCS mixture $S_1N$ (N1-Cyclohexylpropane-1,3-diamine)/ DMCA (N,N-Dimethylcyclohexylamine) which was tested experimentally and exhibited regeneration energy of 2.3 GJ/ton-$CO_2$, and vapor losses and viscosity lower by 10% and 70% than those of other solvents.

In this work, we investigate for the first time the performance of $S_1N$/DMCA in a pilot plant. We perform baseline experiments with 30 wt.% MEA and we test $S_1N$/DMCA at 2M and 3M concentrations. Absorption and stripping are performed at atmospheric pressure, while critical operating parameter data, such as temperature, pressure, pressure drop, liquid levels, gas-liquid flows and gas concentrations are monitored online and registered. Liquid samples are collected periodically for off-line analysis ($CO_2$-loading and amine content titrations). Several KPIs are estimated and monitored for the comparative evaluation of solvent and process performance, such as cyclic capacity, $CO_2$ removal rate, and regeneration energy, under different L/G ratios (mass solvent flow rate/mass flue gas flow).

The results from MEA experiments indicate reliable pilot plant operation, which is in line with literature data [13,14], with an absorber temperature profile in the range of 32-45 °C. MEA further exhibited a maximum cyclic capacity of 0.312 mol $CO_2$/kg solvent, whereas the attained rich loading at steady-state was 11.5% lower than loadings attained from lab-scale equilibrium experiments. This is a clear indication of the satisfactory operation of the pilot plant. For $S_1N$/DMCA, the absorber and stripper columns operated in a range of 31.9-41.7 °C and 84-90 °C, respectively. These results are in-line with the results of Zhang [8], who also performed pilot tests of cyclic amines of similar chemical structures. Both the 2M and 3M solvent mixtures were observed to readily separate into two liquid phases at the phase-separator, placed after the absorber. The $CO_2$ rich phase formed in the 2M solvent mixture case was found to contain lower amounts of $CO_2$ than the 3M case. Therefore, the 3M $S_1N$/DMCA case was selected to proceed in further investigation. At such an amine concentration, the $CO_2$-lean phase had a total amine content of approximately 85 wt.% and enabled approximately a 35% lower solvent flow to the desorber enabling a drastic reduction in the heat of regeneration. Figure 1 further illustrates the cyclic capacity of $S_1N$/DMCA and MEA at different L/G ratios. It is observed that the cyclic capacity of $S_1N/DMCA$ may be up to 3 times higher than that of MEA, while it was further observed that its regeneration energy was up to 59% lower than that of MEA. Additional investigations at different operating conditions will be performed and reported in the conference.

**Figure 1. Cyclic capacity depending on L/G ratio for 3M S1N/DMCA and MEA.**

### Acknowledgements

This research has been co-financed by the European Union and Greek national funds through the Operational Program Competitiveness, Entrepreneurship and Innovation, under the call Research–Create–Innovate, Project T2EΔK-01911. This work has received funding from the European Union's Horizon 2020 research and innovation program under the grant agreement 727503 - ROLINCAP – H2020-LCE-2016-2017/H2020-LCE-2016-RES-CCS-RIA

### References

[1] McMichael, A.J.; Woodruff, R.E.; Hales, S. Climate change and human health: Present and future risks. *Lancet* **2006**, *367*, 859–869, doi:10.1016/S0140-6736(06)68079-3.

[2] Mora, C.; Spirandelli, D.; Franklin, E.C.; Lynham, J.; Kantar, M.B.; Miles, W.; Smith, C.Z.; Freel, K.; Moy, J.; Louis, L. V.; et al. Broad threat to humanity from cumulative climate hazards intensified by greenhouse gas emissions. *Nat. Clim. Chang.* **2018**, *8*, 1062–1071, doi:10.1038/s41558-018-0315-6.

[3] OECD *The Economic Consequences of Climate Change*; OECD Publishing, Paris, 2015;

[4] Institute, G.C. *Global Status of CCS 2019. Targeting climate change.*; 2019;

[5] Simon, F. Meet Europe's two 'most exciting' $CO_2$ capture and storage projects Available online: https://www.euractiv.com/section/energy/news/shell-opens-first-of-its-kind-hydrogen-refuelling-station/.

[6] Bui, M.; Adjiman, C.S.; Bardow, A.; Anthony, E.J.; Boston, A.; Brown, S.; Fennell, P.S.; Fuss, S.; Galindo, A.; Hackett, L.A.; et al. Carbon capture and storage (CCS): The way forward. *Energy Environ. Sci.* **2018**, *11*, 1062–1176, doi:10.1039/c7ee02342a.

[7] Papadopoulos, A.I.; Tzirakis, F.; Tsivintzelis, I.; Seferlis, P. Phase-change solvents and processes for post-combustion $CO_2$ capture: A detailed review. *Ind. Eng. Chem. Res.* **2019**, *58*, 5088–5111, doi:10.1021/acs.iecr.8b06279.

[8] Zhang, J. Study on $CO_2$ capture using thermomorphic biphasic solvents with energy-efficient regeneration., TU-Dortmund, 2013.

[9] Pinto, D.D.D.; Knuutila, H.; Fytianos, G.; Haugen, G.; Mejdell, T.; Svendsen, H.F. $CO_2$ post combustion capture with a phase change solvent. Pilot plant campaign. *Int. J. Greenh. Gas Control* **2014**, *31*, 153–164, doi:10.1016/j.ijggc.2014.10.007.

[10] Raynal, L.; Briot, P.; Dreillard, M.; Broutin, P.; Mangiaracina, A.; Drioli, B.S.; Politi, M.; La Marca, C.; Mertens, J.; Thielens, M.L.; et al. Evaluation of the DMX process for industrial pilot demonstration - methodology and results. *Energy Procedia* **2014**, *63*, 6298–6309, doi:10.1016/j.egypro.2014.11.662.

[11] Hu, L. Methods and systems for deacidizing gaseous mixtures 2010, US 7,718,151 B1.

[12] Papadopoulos, A.I.; Perdomo, F.A.; Tzirakis, F.; Shavalieva, G.; Tsivintzelis, I.; Kazepidis, P.; Nessi, E.; Papadokonstantakis, S.; Seferlis, P.; Galindo, A.; et al. Molecular engineering of sustainable phase-change solvents: From digital design to scaling-up for $CO_2$ capture. *Chem. Eng. J.* **2020**, 127624,

doi:10.1016/j.cej.2020.127624.

[13] Notz, R.; Mangalapally, H.P.; Hasse, H. Post combustion $CO_2$ capture by reactive absorption: Pilot plant description and results of systematic studies with MEA. *Int. J. Greenh. Gas Control* **2012**, *6*, 84–112, doi:10.1016/j.ijggc.2011.11.004.

[14] Thimsen, D.; Maxson, A.; Smith, V.; Cents, T.; Falk-Pedersen, O.; Gorset, O.; Hamborg, E.S. Results from MEA testing at the $CO_2$ technology centre mongstad. Part I: Post-combustion CO2 capture testing methodology. *Energy Procedia* **2014**, *63*, 5938–5958, doi:10.1016/j.egypro.2014.11.630.

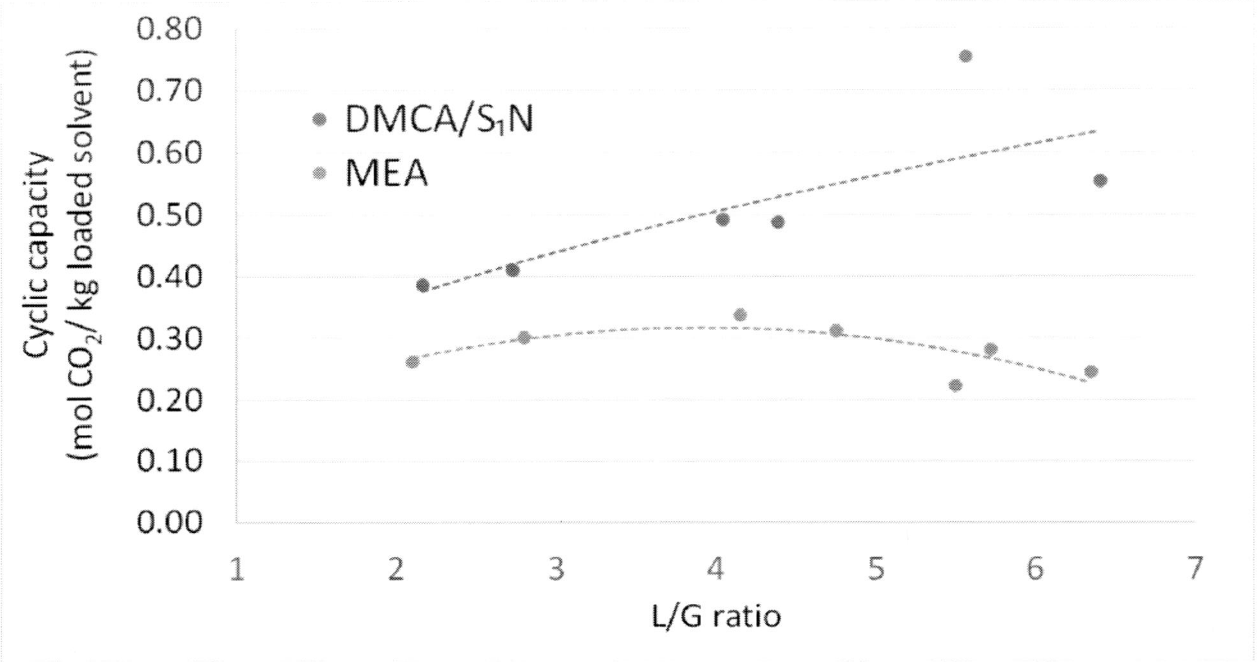

# 630410 3D CFD Simulation of Two-Phase Flow in a Rotating Packed Bed (RPB) for CO2 Capture: Sensitivity of Pressure Drop and Liquid Pool to Operating Conditions

**Abstract:**

Carbon capture processes are very investment intensive, so this is a major driver for technical innovation. Process intensification in the field of gas-liquid contacting may lead to the downsizing of one order of magnitude (if not more) of the structured packing volume required to contact the fumes and the solvent in charge of extracting CO2. Rotating packed beds (RPB) appear to be one of the promising process intensification technologies [1] as it increases both mass transfer and mixing efficiency.

The aim of this study is to contribute to the understanding of the complex two-phase flow in a RPB in order to improve the design of CO2 capture facilities to reduce greenhouse gas emissions, in particular from industrial combustion flue gases.

Steady 3D CFD (Computational Fluid Dynamics) simulations were carried out using ANSYS® FLUENT® with a two-phase liquid-gas Eulerian approach. The packing is modeled as an isotropic porous bed zone. The liquid phase is assumed to be described as mono-sized droplets using an Euler-Euler modelling.

The previously developed CFD model [2] seems to reproduce properly the global features of the flow. Nevertheless, the range of the operating conditions has to be extended. In this work, the rotation speed (w), as well as the liquid mass flow rate ($Q_L$) and physical properties were varied to investigate the sensitivity of flow field, pressure drop and liquid pool at the bottom of the RPB (Figures 1 and 2).

Besides, different geometries were also compared.

***References***

[1] "Mass Transfer in a Rotating Packed Bed: A Critical Review", Z. Wang, T. Yang, Z. Liu, S. Wang, Y. Gao, M. Wu, Chemical Engineering & Processing: Process Intensification, 139, 2019.

[2] P. Béard, V. Penin, P. Alix, "Towards 3D CFD simulation of two-phase flow in a rotating packed bed (RPB) for CO2 capture", CHISA (International Congress of Chemical and Process Engineering), virtually, 2021.

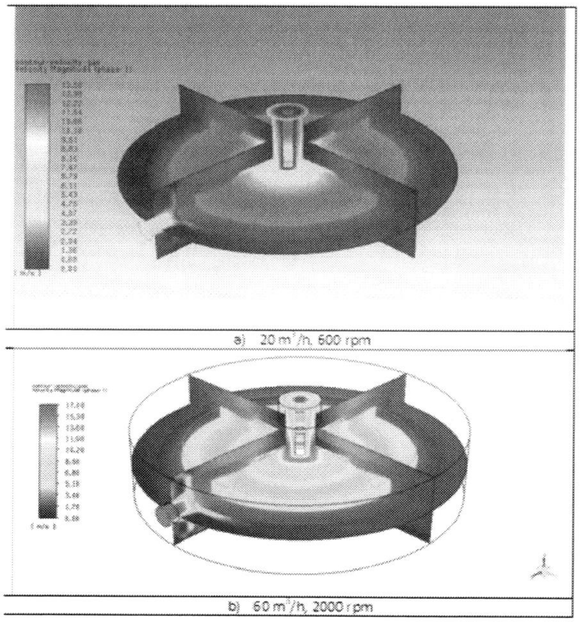

a) 20 m$^3$/h, 600 rpm

b) 60 m$^3$/h, 2000 rpm

Figure 1 : Gas velocity fields in 3 planes in the RPB and in gas outlet for different operating conditions.

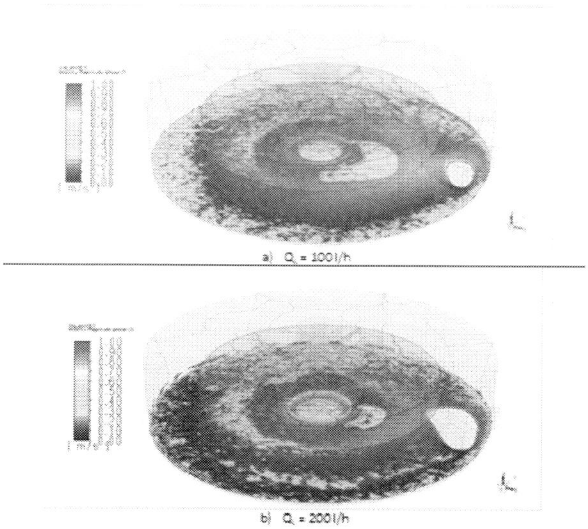

a) $Q_L$ = 100l/h

b) $Q_L$ = 200l/h

Figure 2 : Influence of liquid mass flow rate on liquid pool (iso-surface of liquid volume fraction 0.5) and liquid velocities on the pool surface in the RPB (60 m$^3$/h, 2000 rpm).

# 630389 Optimal Selection of Solvent and Contactor Technology for Micro-Encapsulated Carbon Sorbent (MECS) Based $CO_2$ Capture

**Abstract:**

Carbon capture utilization and storage (CCUS) is believed to play a prominent role in combating climate change and global warming issues. Many novel technologies are being developed to make $CO_2$ capture an economically viable piece of the net-zero carbon framework. There are several potential solvents such as ionic liquids, carbonate solutions, and phase change solvents that have favorable energetics and can achieve high loading with promise of significantly improving the economics of $CO_2$ capture processes. However, these solvents can suffer from high viscosity and/or solid reaction products that can lead to transport difficulties, clogging of the contactor, and the use of distillation towers impractical. Microencapsulation of such solvents inside a polymer capsule can help alleviate some of these operational difficulties. Microencapsulation of carbon sorbents (MECS) is a technology proposed by Vericella et.al. for $CO_2$ capture applications [1]. Microcapsules containing the solvent can be produced with diameters ranging from 100-600 microns. The small size of these microcapsules results in a high specific surface area per unit volume, which can enhance mass and heat transfer rates. A few studies [2-4] have reported analysis of MECS in a fixed bed, and fluidized bed reactor configuration for a particular solvent (mainly carbonate solutions and MEA). As the choice of microencapsulated solvents and the type of contactor both have strong impact on the economic feasibility of this technology, detailed modeling and techno-economic analysis is required to properly evaluate this technology.

The present work evaluates different solvents and reactor configuration for MECS using process modeling and techno-economic analysis. The encapsulated solvents considered in this study are $Na_2CO_3$, ionic liquids, and piperazine. Required thermodynamic, reaction kinetics, and physical properties models of the solvents are developed based on the data obtained from the open literature. First, a rigorous capsule model is developed with an underlying solvent chemistry model that describes heat and mass transfer in the microcapsule. The results from the capsule model are validated with experimental data for the $Na_2CO_3$solvent [4]. Two contactor technologies are evaluated: a fixed bed and a moving bed. These reactor models are 1-D non-isothermal, first principles-based models with appropriate mass transfer, hydro-dynamic, and momentum sub-models. The multi-scale reactor models incorporating the capsule model and embedded heat exchangers are implemented in Aspen Custom Modeler (ACM) where the method of lines is used to solve the resultant system of partial differential equations. The models simulate both absorption and regeneration stages of the process and study the impact of key design and operating parameters on the performance of MECS technology.

Techno-economic analysis of these reactor types operating with different solvents is carried out by computing their Equivalent Annual Operating Cost (EAOC) and comparing them with the traditional MEA technology [5]. The capital cost of the reactor, compressors, and heat exchangers is estimated using Aspen Process Economic Analyzer (APEA) along with custom correlations. Operating costs include the cooling and heating utilities and compression cost for the flue gas, if any. Each combination of reactor type and solvent type is optimized for a fair comparison. It is observed that both reactor type and specific solvent have large impact on the economics of MECS technology.

## References


1. Vericella, J. J. et al. "Encapsulated liquid sorbents for carbon dioxide capture", Nature Communications, 6:6124, 1-7, 2015.
2. Anggit Raksajati, et al, "Techno-economic Evaluation of $CO_2$ Capture from Flue Gases Using Encapsulated Solvent", Industrial & Engineering Chemistry Research 1604-1620, 56 (6), 2017.
3. Hornbostel, et al, Packed and fluidized bed absorber modeling for carbon capture with micro-encapsulated sodium carbonate solution, Applied Energy 1192-1204, 2019.
4. Kotamreddy, et al., Process Modeling and Techno-Economic Analysis of a CO2 Capture Process Using Fixed Bed Reactors with a Microencapsulated Solvent, Energy & Fuels 2019 33 (8), 7534-7549 DOI: 10.1021/acs.energyfuels.9b01255.
5. T, Cost and Performance Baseline for Fossil Energy Plants Volume 1. 2015. DOI: DOE/NETL-2015/1723.

# 632027 Surface Degradation of Li-Ion Battery Cathodes from First Principles

**Abstract:**

For the transition from traditional fossil fuels to environmentally friendly renewable energy, lithium-ion batteries based on nickel- and lithium-rich cathodes could potentially provide the required high energy densities as an efficient energy storage system. However, Ni-rich high-energy-density Li-ion battery cathodes are prone to thermal degradation, which leads to safety concerns and hinders commercial adoption. Degradation typically occurs at the particle surface and involves the release of oxygen gas. To identify stable Ni-rich cathode compositions or surface coatings that can prevent undesired phase transformation, an understanding of the fundamental degradation processes on the atomic scale is needed.

Here, we apply first principles atomistic modeling to shed light on surface reactivity, providing information that is complementary to experimental characterization. The mechanism of the self-reductive decomposition of $LiNiO_2$ is determined using automated density-functional theory calculations. We construct first principles phase diagrams that map out stable surface phases as a function of the state of charge and the temperature. The results obtained from this analysis allow deducing the degradation mechanism and offer guidance for the development of degradation-stable cathode compositions and improved coating materials.

# 632417 Liquid Structure in Water-in-Salt Electrolytes

**Abstract:**

Lithium ion ($Li^+$) batteries were first commercialized about three decades ago and have been widely used since then, especially in mobile electronics. However, fire and explosive accidents in recent years have raised safety concerns regarding the flammable organic-based electrolyte solutions usually used in these batteries. Organic electrolytes also have cost and environmental concerns and complicate battery recycling. Aqueous electrolytes offer a safer, more environmentally benign alternative. Unfortunately, the narrow electrochemical stability window of water (1.23 V) limits energy density of a battery and hence their appeal for commercial applications. The recent discovery of the water-in-salt (WIS) electrolyte concept provides an exciting and promising new route for the design of aqueous electrolyte batteries. However, fundamental knowledge of the solvation structure of these electrolytes, which is a prerequisite for the understanding ionic transport and electrochemical properties, remains limited.

In this work, two WIS systems were studied using combined X-ray scattering, X-ray absorption, FTIR experiments and classical molecular dynamics (MD) simulations. One WIS is a 20 molal (m) LiTFSI aqueous solution and the other consists of 1 m $Zn(TFSI)_2$ and 20 m LiTFSI. Classical MD simulations are validated against the experiments. Both experiments and simulations demonstrate that in these highly concentrated WIS electrolytes, the water network is disrupted and the majority of water molecules exist in the form of isolated monomers, clusters or small aggregates with chain-like configurations. On the other hand, $TFSI^-$ anions are connected to each other and form a network. This description is fundamentally different from those proposed in earlier studies, which suggested that a water-rich domain and a $TFSI^-$-rich domain coexist in the liquids. In addition, for the WIS doped with 1 m $Zn(TFSI)_2$, we found that $Zn^{2+}$ ions are solvated by six waters in their first solvation shell and $TFSI^-$ anions are completely excluded although $TFSI^-$ concentration is as high as 22 m. This solvation structure is also fundamentally different from the picture that $Zn^{2+}$ ions are completely coordinated by three $TFSI^-$ anions proposed in a previous study. The new pictures of the liquid structures revealed in the current study suggest that new mechanisms are required to understand the transport and electrochemical properties of these WIS electrolytes.

# 630725 Performance and Durability of Vapor-Fed Unitized Regenerative Fuel Cells

**Abstract:**

As the global demand for electrical energy increases and the drive toward carbon-free electricity generation intensifies, it is necessary to develop energy storage devices to compensate for the inherent intermittencies of renewable energy sources, such as wind and solar. One such device is a unitized regenerative fuel cell (URFC), which is effectively a hydrogen battery. The URFC has two modes of operation: water electrolysis mode, which utilizes renewable electrons to produce carbon-free hydrogen, and fuel cell mode, which generates electrical energy for on-demand electricity. The URFC system thus operates with both liquid and gas phases, which complicates water management within the cell, increases transition time between modes, and produces wet hydrogen gas that will need to be dried downstream.

In this study, a URFC using a water vapor feed (vapor-URFC) in electrolysis mode is presented as an alternative option. Using water vapor allows the vapor-URFC to operate exclusively in gas phase, which reduces transition times between the two operating modes. We compare the performance of the vapor-URFC to a traditional URFC operating in the constant gas configuration, in which hydrogen reactions (oxidation and evolution) occur at one electrode, and oxygen reactions (reduction and evolution) occur at the other electrode. Performance and durability of the cell are assessed by measuring polarization curves and performing accelerated stress tests. Preliminary results show that a vapor-URFC is able to obtain current densities of 2 $A/cm^2$ at 0.2 V and at 2 V during fuel cell and electrolysis operation, respectively. A vapor-URFC is also able to achieve a round trip efficiency of 60% at 1 $A/cm^2$. Furthermore, we present a parametric study of different ion exchange membranes and porous transport layers to determine the optimal architecture for the vapor-URFC cell. Finally, we analyze the performance of the vapor-URFC over a range of relative humidities and using varying carrier gases to demonstrate the viability of this device in a variety of operating conditions.

# 631598 Thin-Film Polymer Coatings By Molecular Layer Deposition for Advanced Energy Storage Electrodes


**Abstract:**

To provide both high power density and high energy density in a single electrochemical energy storage device, it is appealing to generate electrodes with a thin-film redox-active coating on a high-surface-area conductive support. Atomic layer deposition (ALD) is a proven technique to achieve this desired electrode configuration. ALD employs sequential, gas-phase surface reactions to form uniform, thin-film coatings of redox-active materials onto high-surface-area, three-dimensional substrates. Unfortunately, the ALD precursors needed to grow inorganic redox-active materials are too expensive to be viable for depositing thin-film coatings of redox-active material in commercial energy-storage devices. In this work, we report on efforts in our group to form electrodes coated with thin-film redox-active polymers using molecular layer deposition (MLD). MLD is a molecular analog to ALD and employs low-cost organic precursors. We describe experimental studies on new MLD chemistries that we use to form redox-active polymer thin-films based on conjugated amines. We examine the infiltration of these MLD coating chemistries into porous electrodes, and measure the electrochemical properties of the resulting polymer films and electrodes.


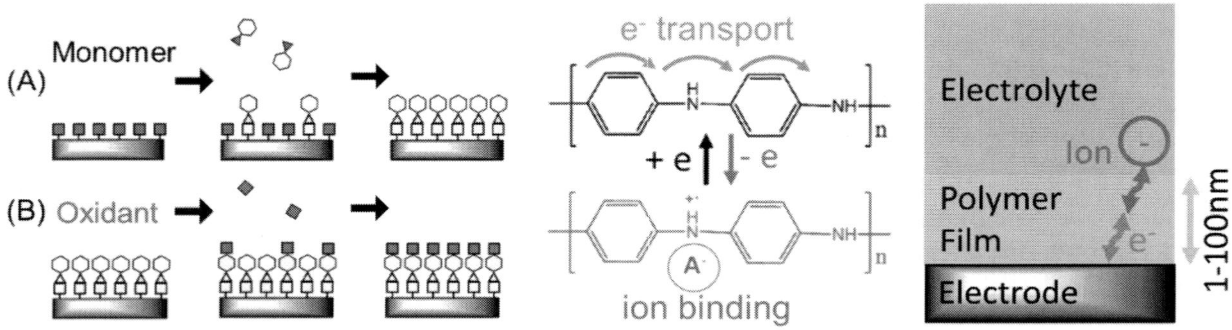

# AUTHOR INDEX